Microcomputer Quantum Mechanics

Microcomputer

Quantum Mechanics

Second edition

J P Killingbeck, DSc

Department of Physics,
University of Hull

Adam Hilger Ltd, Bristol and Boston

British Library Cataloguing in Publication Data
Killingbeck, John
   Microcomputer quantum mechanics.
   1. Quantum theory — Data processing
   2. Microprocessors
   I. Title
   530.1'2   QC176.96
   ISBN 0-85274-803-5

First printed 1983
Second edition 1985

Consultant Editor: **Professor M H Rogers**, Department of Computer Science, University of Bristol

Published by Adam Hilger Ltd
Techno House, Redcliffe Way, Bristol BS1 6NX, England
PO Box 230, Accord, MA 02018, USA

Printed in Great Britain by J W Arrowsmith Ltd, Bristol

Contents

Preface to the second edition

I am pleased that the book has been so well received, both by readers and by reviewers. Since I tried to stress the mathematical as well as the computing side of my subject, I am also pleased to learn that some readers have found my earlier book useful (*Techniques of Applied Quantum Mechanics* is available from Adam Hilger Ltd). For this new edition I have corrected some misprints, changed Appendix 2 and written a new chapter which describes recent developments. My initial intention was to extend §1.4 to mention several new microcomputers, after a proper test to see how they handle my programs. I accordingly tried to enlist the help of some computer companies. I am sorry to say that Texas Instruments and CBM proved to be rather unhelpful and IBM also seemed to have little interest in my work. By contrast, it is a pleasure to thank Sinclair Research for their continuing support of my work on microcomputer methods. Most of the calculations in the new chapter, as well as in my recent papers, have been carried out on a Sinclair Spectrum microcomputer, to emphasise that good scientific work can be done on low-priced machines if the appropriate mathematical methods are employed.

Preface to the first edition

Much of this book is about the wise use of microcomputers in scientific work, and so should be of interest to a wide group of students and research workers. The first few chapters, dealing with general ways of applying and testing micro-computers, may well be of value to teachers who are beginning to use them in school work. To give the work some focus, however, I have taken as my subject in later chapters the use of microcomputers in simple mechanics, particularly quantum mechanics, so that these chapters are best suited to a reader who has some knowledge of quantum mechanics. For example, they would constitute a useful 'numerical applications' course to run in parallel with a course on the basic principles of quantum mechanics. It is my belief that a book on computing gains force if it actually shows the way to apply computing methods to some real problems. Many students nowadays take compulsory courses on 'computing'; they often end up knowing a language but with nothing to say (a modern variant of being all dressed up with nowhere to go). What I try to illustrate in this book is the way in which computation is usually integrated with theoretical analysis as part of a unified attack on a scientific problem. I do this by giving many case studies in which I set out stage by stage the way in which various problems could be handled.

I have made an attempt to keep the mathematics as simple as possible, in the sense that the reader will not need a great knowledge of mathematical facts to follow the work. It is the application of the simple principles in diverse circumstances which I emphasise. For example, the simple formula for an iterative inverse and the formula for the first-order energy shift are applied repeatedly in many different connections throughout the book, and the use of recurrence relations and of Richardson extrapolation is a persistent theme. All the mathematical notation is fairly standard, although I vary it (e.g. using y' or Dy) at some points to keep the equations neat.

Although I chose topics in the light of my personal experience and interests, it turned out that my choice was in some ways complementary to that made by

J C Nash in his book *Compact Numerical Methods for Computers* (Bristol: Adam Hilger 1979). For example, Nash gives a more detailed study of matrix eigenvector calculations than I do, but I look in more detail than he does at problems of numerical integration and of eigenvalue calculation for differential equations. Taken together, his book and mine cover a wide range of material which is of value for users of small computers.

Many of my own programs are used thoughout the book, although I sometimes refer to recent sources where a tested BASIC program has already been published. A few specific machines are used to show how the various methods work, but I must emphasise that what matters is the simple structure of the programs, which makes them adaptable for almost any microcomputer. (When modified, the programs which use large arrays, and which I tried on a CBM Pet, will work on a Sinclair ZX-81 when it is equipped with one of the low-priced 16K RAMs which have recently become available.)

Throughout the text I provide many examples and worked exercises to help the reader develop both analytical and numerical skills, and I give references to books and papers in which further details about particular topics can be found. I do not assume that the reader is adept at BASIC program writing but I do assume that he has studied the handbook for his particular microcomputer with care. He can then judge how best to carry out for his machine any general procedures which I describe, although almost every machine will resemble at least one of my four specimen ones.

Since I concentrate on using simple computer methods to calculate various quantities, at some points I have had to assume that the reader has some knowledge of basic quantum mechanics. Readers who wish to look more deeply into the theory behind the use of variational methods, perturbation theory and group theory in quantum mechanics will find an overall survey in my earlier book *Techniques of Applied Quantum Mechanics*. That book, originally published by Butterworths, is now available from Adam Hilger.

There are two acknowledgments which I am pleased to make. I thank Neville Goodman for cajoling me into writing this book for Adam Hilger. He has the inestimable gift of being a professional amongst publishers and a friend amongst authors. That I managed to do the job in anything like reasonable time was due to the unstinting help of Margaret Bowen, who has my respect and admiration for her speedy and efficient preparation of the manuscript.

1 Microcomputers and BASIC

1.1 What is a microcomputer?

In the current state of growth in the computer hardware industry more and more power is being packed into small computers. It is not always clear whether the conversational prefixes 'mini' and 'micro' refer to physical size or to computing 'size' (in kilobytes of RAM) or, indeed, whether they can be used interchangeably. For the purposes of this book I arbitrarily take a microcomputer to be a computer with 8 kilobytes or less of RAM. This is not a dogmatic definition of what a microcomputer *is*, but rather a statement about the kind of machine which can handle the simple methods described in this book (i.e. a statement of my terms of reference). Many of the calculations can actually be handled in a memory space of order 1K; to make sure that this is so they have been designed and tested using some typical machines. These are: a Texas Instruments TI-58 programmable calculator, a Sharp PC-1211 pocket computer, a Sinclair ZX-81 computer of the basic 1K type and a CBM Pet computer of 8K type. The last three machines will accept programs written in the BASIC language, which is widely used in small computers. ALGOL and FORTRAN are often used by scientists working with mainframe computers, and several people have championed languages such as PASCAL and COMAL 80 as being in some ways preferable to BASIC for small computers. Although some computers (e.g. the Apple and the Pet) can be obtained in versions which use other languages, it is still the case that most purchasers of an 'off the shelf' home computer will find it already set up to use BASIC. Accordingly, I have used BASIC for much of the material in this book, although it is not difficult to convert my programs to other languages once the underlying structure of the algorithms has been understood. Indeed, even the dialects of BASIC differ a little from one machine to another, as pointed out in the excellent guide by Alcock [1].

Programmable calculators such as the TI-58 which have conditional jump and subroutine facilities are classed by me as computers [2]. I take the important threshold between calculator and computer to be crossed when a machine's stored program can include instructions which control the flow of the calculation by means of conditional jumps, loops, etc. Of course, most calculators work with *numbered* stores rather than *named* ones. For example, in a computer using BASIC we could have a calculation using variables X, Y and Z and put in the middle of the program a line such as

LET $Z = X + Y$

The computer would get X and Y and store the sum Z without us knowing in which locations these quantities are kept; it would do the book-keeping for us. To do this calculation in a TI-58 program we would have to decide to keep (say) X in store 1, Y in store 2 and Z in store 3 and would have to remember this carefully. To set $Z = X + Y$ we would use the instructions

RCL1 + RCL2 = STO 3

with the book-keeping explicitly done by the human programmer. Despite this extra problem and also its slower speed as compared to most computers the TI-58 is a worthwhile instrument for several of the calculations in this book, since it works to 13 digits accuracy. This is greater accuracy than most computers achieve when using BASIC. For some types of iterative calculation, in which the required number is approached more and more closely on each cycle of the calculation, it is sometimes useful to gang together (say) the TI-58 and the Pet. The Pet quickly does many cycles to get a good answer and the slower calculator only needs to do the last cycle to give a very accurate result.

In a comparatively short book I cannot do everything. I have no doubt that a machine language expert would be able to group together the 8-bit bytes in the ZX-81 or the Pet so as to produce multiple precision arithmetic and also to increase calculating speeds. I have thought about this possibility while writing this book and decided not to be drawn into that area, but rather to concentrate on the way in which the *analysis of the underlying theory* can help to improve the speed and accuracy of calculations. Most of the calculations which I describe involve many multiplications and divisions; these are rather tedious to write in machine code since most chips have only addition and subtraction as basic instructions. Accordingly, I declined to take what for me would be a lengthy detour, but I am sure that a valuable book (in some senses complementary to this one) is waiting to be written by a machine code expert who is also a physicist. One thing which programmable calculators cannot do but BASIC language computers *can* do is to handle strings of symbols (e.g. sorting words into alphabetic order). This difference is not of much importance for the material of this book, since I concern myself mainly with numerical algorithms rather than information and data handling tasks.

1.2 Interaction and iteration

Some scientists who use computers take the view that a 'good' program is an entirely automatic one. The operator has to supply some necessary data, but thereafter all the calculations and decisions are controlled by the program; the operator can go off for a round of golf and pick up the results later. Nowadays there are many library programs which operate automatically in this way on mainframe computers (and indeed on microcomputers), but I am sure that there is still plenty of scope for interactive computing on microcomputers. By their nature library programs are usually run by people who did not design them and so the limits of tolerance (of the program and of the people) are of crucial importance. For example, several software designers at a conference which I attended [3] noted that scientists were wrongly using library programs for differential equations in circumstances where the underlying theory behind the programs shows them to be of doubtful accuracy. (I was reminded of the well known saying, usually employed in another connection: designed by a genius, to be run by an idiot.) In my own area of quantum mechanics I have often noticed people attacking simple problems by using matrix methods on large computers, apparently for no better reason than that the library programs for handling large matrices 'are there' (and thus represent a solution looking for a problem). With a little thought it is sometimes possible to get better results using a simple non-matrix method on an electronic calculator. However, as was pointed out to me once by someone in the computer industry, in many institutions a man's importance I is assumed to be proportional to the computer time T which he uses. For my own part I think that a formula of form $I = AT + BT^{-1}$ is more appropriate; the T^{-1} term is an 'ingenuity term'.

In this book, then, I want to deal with calculations which can be done interactively on microcomputers. The operator will from time to time stop and restart the calculation and may need to monitor some output data. If the calculation looks 'wrong' then he can insert at will an output instruction at any stage of the program to monitor intermediate results, and can later wipe out that instruction when the debugging is completed. Many scientists actually feel surer about a calculation if they can trace it through and control it in this way. In physics there are many problems (e.g. large-atom Hartree–Fock calculations, analysis of x-ray crystallographic data) which really *need* large computers, and we may safely leave these as the proper business of the large machines. However, there are many smaller but important calculations which can be handled in a more intimate interactive mode on microcomputers such as the four typical ones listed in §1. In interactive computing the aim is to combine the operator's experience and judgment (which are difficult to capture in an automatic program) with the computer's great calculating speed (which, Zerah Colburn and Carl Friedrich

Gauss [4] excepted, is beyond human capability). In this sense, then, the human operator plus the microcomputer forms 'the computer'. This point of view, which I have outlined elsewhere [2] has been with me ever since, years ago, I read a science fiction story in which the main character, frustrated at trying to design a computer which can handle the ambiguities of ordinary language, builds himself into the machine [5].

The problems which I treat in this book arise from classical and quantum mechanics; these branches of physics provide sufficient problems to represent an interesting sample of microcomputer programming and numerical analysis, so that the topics which I discuss have ramifications throughout much of science and mathematics. My main emphasis is on taking simple mathematics as far as it will go, or, as I sometimes say, taking it seriously. The point is that there is some tendency in the current scientific literature to regard only sophisticated-looking mathematics as 'important'. To quote but one of many examples: I recently saw a paper of ten or so pages, full of very clever contour integrations, which it would take a reader weeks to unravel. The end product was a number, of relevance in quantum mechanics, accurate to about five per cent. It turned out to be possible to get the number to 1 part in 10^6 by a simple calculator trick using two pages and no contour integrals. If a scientific writer believes, as I do, that the aim of writing is to communicate and, in particular, to increase the knowledge and scientific power of many readers, then obviously his best procedure is to use simple mathematics and short arguments. The *deployment* of the mathematics may be original, but the palace should be built of ordinary bricks. (Most physicists remember that Weyl accused Dirac of secretly using group theory in a supposedly elementary lecture; Dirac replied that his approach had not needed any *previous knowledge* of group theory [6].)

If a piece of mathematics is to be translated into an algorithm which will fit into 1K or so of microcomputer memory then it cannot in any case be too complicated or lengthy, although it might be quite 'clever' in the sense that it takes a careful analysis to see that the required result can be reached so simply (§§6.4 and 7.4 provide examples of this). One common way to make a little go a long way is to translate a calculation into a form in which it can be accomplished using an iterative or a recursive algorithm. Only one cycle of the iteration needs to be programmed, giving a short program, and the calculation simply repeats the cycle many times. The operator can see by eye when the iteration has converged if the calculation is done in an interactive manner. Here again, some purists would say that stopping an iterative process should be done automatically, with the computer stopping when two successive iterates agree to a specified accuracy. However, if we have an equation with roots, say 0.0011 and 8315.2, do we specify 'stop when the two estimates differ by less than 10^{-4}' or 'stop when the results differ by less than 1 part in 10^4'? In some cases an early 'pseudo-limit' can be reached while the actually required limit is only reached after a long run. Clearly then it is not always obvious how to write an automatic set of program instructions to handle even this simple END command. Of course, a sufficiently long program could take care of most eventualities, but it would take longer to write and it would involve several logical decision steps, which slow down the running speed of the program. It is often better to save ourselves doubts about the validity of the results by simply doing the calculation interactively and monitoring the calculation on the display screen. Machines such as the ZX-81 and the Pet have print position control instructions which can be included in the program, so that the output value of the required quantity X, say, on each iterative cycle can be printed at a fixed screen position. As the values X_1, X_2, etc converge to the limit the leading digits of the displayed number freeze; the later digits whirl over and freeze one by one into their final converged values. Since only one screen line is used by the whole calculation of many iterative cycles, the results of several successive calculations can be displayed on the screen at one time. In the cases of the TI-58 and the Sharp PC-1211 only one line is visible at a time in any case, although the PC-1211 can put more than one number on that line; it could for example show X_n and X_{n+1} side by side. Both machines have a PAUSE instruction which will hold the calculation up while it displays the required number for about a second. On the TI-58 the pauses can be strung together to display the number for as long as required (e.g. so that it can be copied down onto a notepad).

1.3 Theory in action

In the study of quantum mechanics the use of a microcomputer is valuable in both teaching and research. At the teaching level it makes it possible for students to see how methods such as variational theory or perturbation theory turn out when they are used numerically and it makes clear which bits of the formal textbook algebra can be easily put into practice when it actually comes to putting numbers in the equations. It also encourages a more flexible approach to the use of mathematical equations themselves by leading the student to see that the 'best' form for expressing an equation or its solution may not be the same in analytical and numerical work. Thus, rather than trying to find a complicated explicit formula for the value of an integral it may be worthwhile to simply work it out numerically. There is a reverse side to the coin, of course: sometimes if we do know an exact (and exactly calculable) analytic solution to a problem we can use it as a test problem to check whether a proposed numerical procedure on the computer is sufficiently accurate or stable. This approach can be used in a valuable two-way feedback process. For example, by working out integrals of various simple functions using the midpoint integration rule, some of my students in a first year university class discovered that for small integration strip widths h the value obtained using the midpoint rule differed from the exact analytical value by an amount proportional to h^2. Having discovered this rule from a sequence of test problems, they could then use it to work out accurate values of integrals which could not be done analytically. This ability to proceed from the known to the unknown in gradual steps is an important part of learning and the microcomputer can be used to aid the process. In a sense it allows a mixture of theory and experiment, albeit in a much smaller world than that encompassed by the whole of physics.

In later chapters I try to illustrate how the use of empirical 'try outs' on a microcomputer can suggest interesting topics for theoretical investigation and I also point out repeatedly that very often a deeper understanding of the mathematics behind a problem can lead to the formulation of better numerical programs for it. In many cases physicists encounter numerical problems as part of some *total* problem which they are handling, and so whenever possible I proceed by using case studies which stress the way in which algebraic and numerical skills usually interlock in actual calculations. This integrated way of looking at the subject was stressed by Fox and Mayers [7] in their admirable book and I agree entirely with their view that getting the analysis right first is an important part of the process before the computer program is written. In the following recipe every ingredient is important. First the problem must be 'caught', i.e. captured in some clearly defined form; very often research problems start as ill defined worries, which only crystallise out as a clear mathematical task after much intuitive trial and error work. Second, the relevant algebra must be worked out with a clear view of what is known and what is to be found, although these input and output requirements may not be formulated numerically in the early stages. (We might say, for example, 'this thing can't get out of here', which

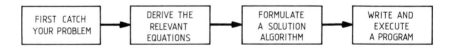

later on becomes the boundary condition $\psi = 0$.) At the third and fourth stages, in which a solution algorithm and a program are formulated, it is almost impossible to avoid interplay between the stages, since the algorithm used may have to be adapted to the capabilities of the computer which is to be used to do the numerical work. In particular, if we wish to keep the program length short we may well want to construct an algorithm which uses an iterative or recursive procedure, even though there exist other types of algorithm which would be theoretically adequate to provide a solution. Even amongst possible iterative algorithms some may be more efficient than others when we take into account the number of significant digits and the operating speed of the computer. In such a case, of course, it may be possible to attack some test problems to get an experimental feel for the relative merits of two proposed algorithms.

Most physicists would regard the semi-empirical approach outlined so far as 'the obvious thing to do', and I agree, although I can see the validity of the hard line numerical analysts' view that such an approach is not respectable unless backed up by rigorous mathematics. One reason why their strictures are relevant is that any worker who has not been involved in the total formulation of the problem may not fully see the implications of changing some of the circumstances; the misuse of library programs which I mentioned earlier is a case in point. Another reason is related to the age-old problem of inductive logic: just because a method works for a few trial cases it doesn't follow that we can rely on it. Since I suspect that most physicists are inclined to pay little heed to such purist admonitions I would like to play devil's advocate just for a moment by citing an example from my own experience. Fairly frequently students who own programmable calculators are initially reluctant to write their own experimental programs to do numerical integration because, as they put it, 'my calculator already has a module program in it to do integrals'. Closer investigation reveals that neither in the calculator handbook nor in the student's head is there any information about how varying the strip width h affects the accuracy of the result. (It is, of course, precisely this information which I am trying to encourage the student to discover.) If the student uses his module program, usually a Simpson's rule one, to integrate x, x^2 and x^3, he will get three results which are exactly right, except perhaps for a tiny rounding error. I have known students (and, indeed, a college lecturer) to reach the unwarranted conclusion that the module program gives exact integrals. In fact, for small h, theory shows that the error varies as βh^4, where β happens to be zero for integrands of form x^n with $n = 0, 1, 2, 3$.

In the kind of investigations outlined above it is clear that at least a little theoretical analysis is needed to help us monitor the empirical investigation. The only moral to be drawn from all this is a fairly general one; always devise as many cross-checks as possible to avoid error, and always be flexible-minded enough to try something new even when the orthodox dogma forbids it. One useful feature of the microcomputer is that it lets a research worker give a quick try-out even to 'silly' ideas which years ago he would have discarded as of little promise. Some of the very simple methods in this book arose in that way; as a physicist who needed some answers I simply proceeded by devising and testing what seemed to me the most simple and direct ways to solve my problems. Some of the methods were so simple that the 'official' numerical analysts, acting as referees for scientific journals, assured me that they obviously could not work. By now, of course, there is sufficient accumulated evidence to excuse my heresies. (There can only be seven planets. But there are nine. So much the worse for the facts.) Some of my early programs were written for programmable calculators with very few memories; this forced me to concentrate on compact ways of formulating the calculations, whereas a larger machine will permit a worker to become lazy in such matters. The stimulation of ingenuity which such constraints produce has its analogue in the arts. Richard Wilbur [8] speaks of the way in which the constraint of using a given verse form can encourage a poet's creativity.

1.4   Varieties of BASIC

It is quite easy to do complicated calculations while using only a few essential parts of the BASIC language, just as one can get by with only a limited selection from the FORTRAN vocabulary [9] or, indeed, as shown by the work of C K Ogden or C Duff, one can manage with a fairly limited vocabulary (a thousand words or so) in several human languages [10]. My main concern in this book is to make the structure of the calculations as simple as possible, so I shall try to make do with the kind of BASIC which can be almost 'picked up as we go along'. However, many people meet BASIC for the first time when they use some *particular* microcomputer, and each manufacturer tends to have his own variant of the language, so that what works on one machine might not work on another. In such cases, the instruction manual for the particular machine should be consulted, but a book such as that by Alcock [1] will provide a useful survey of most of the possible variations in BASIC which can be encountered. The most obvious difference between machines is that some of them need the prefix LET and others do not. To repeat my example of §1.1, in the statement

$$\text{LET } Z = X + Y$$

the word LET is needed on the ZX-81 microcomputer, but not on the PC-1211 or the Pet. (However, the latter machines will work if you put the LET in, whereas the former will give a syntax error signal if the LET is omitted.)

Almost all of the currently available microcomputers have a set of error signals which they give to tell the operator when his program is unacceptable, and they usually indicate the particular program lines in which the errors appear. This is very helpful to someone learning to use a new machine, although it can sometimes result in a kind of guessing game between the microcomputer and the operator. When some offending line seems to be full of syntax errors (and some machines have a few subtle ones not mentioned in the handbook) it sometimes pays to cut your losses and start again, re-ordering the operations, just as in writing it may be better to reformulate a long sentence such as this one when it cannot be patched up easily. Usually the simple way to avoid or localise errors is to use what I term the Hemingway style [11], breaking a calculation up into short statements rather than writing very long expressions with lots of nested brackets and arithmetic operations. If a lengthy line contains some subtle error this breaking up process, giving each fragment its own line number, will help to pinpoint the error. When this has been located and corrected the original longer line can be reconstructed if required; for example, it may help in the visual display to keep the lines fairly full so that the whole program can be seen at once on the screen.

me emphasise again (syntax error). Let me emphasise again that I deal mainly with numerical calculations in later chapters, so that the remarkable data handling capabilities of the various microcomputers are not given much space. (They would presumably form part of the hypothetical machine code book which I propose in §1.1.) With this restriction of my terms of reference in mind, I have drawn up a list of a few of the useful features which typical microcomputers have. To show the variety of facilities available I have also compiled a table to show how four particular machines exemplify these features, together with a brief survey of those properties which give each machine its own special 'personality'.

List of Features
1. Can store a program (on magnetic tape or in a permanent solid state memory).
2. Can accept program statements both with and without LET.
3. Can accept several statements per line, with separating colons (:).
4. Single key facility available for all commands (RUN, PRINT, LIST etc).
5. Availability of common scientific functions (EXP, COS etc).
6. Can accept user-defined functions.
7. BODMAS arithmetic, e.g. $3 + 4 * 2$ is evaluated as $3 + (4 * 2)$.
8. Overflow stops program execution and gives error signal.
9. Works to a limited number of digits.
10. Has GOTO, conditional jump and subroutine facilities.
11. Programs can be RUN starting at any line.
12. Variables can be changed manually between STOP and CONT instructions.
13. Fixed position printing possible to give a static display for iterative processes.
14. Can work with matrix arrays specified in the form, e.g. $M(I, J)$.

	1	2	3	4	5	6	7	8	9	10	11	12	13	14
TI-58C	√	—	—	—	√	—	√	IE100	13	√	√	√	—	×
PC-1211	√	√	√	√	√	√	√	IE100	10	√	√	√	—	×
ZX-81	√	×	×	√	√	×	√	IE38	9	√	√	√	√	√
Pet	√	√	√	×	√	√	√	IE38	10	√	√	√	√	√

Texas Instruments TI-58C

The TI-58C and TI-58 differ only in that the 58C has a continuous memory which can retain a program. The calculator has a flexible memory, which contains up to 480 program steps or up to 60 stores, the two being interconvertible in an 8:1 ratio. It has an internal module containing ready-made programs for a variety of calculations; these are called by an appropriate code number. It is possible to act on numbers while they are in the stores, e.g. 15 SUM 1 SUM 2 adds 15 to the numbers in stores 1 and 2. A program can be listed one step at a time while insertions or modifications are made. Conditional jumps are made by testing the display number against the contents of a special t register using an $x \geqslant t$ test to decide whether or not to make the jump. The possible jump destinations can be specified by absolute numerical step locations or by letters A, B, C etc. Errors such as a request for $\sqrt{-5}$ produce a flashing display of $\sqrt{5}$ rather than an error signal. (Asking for SQR (-5) on most microcomputers gives a halt and an error signal.) Although the display shows only ten digits, the full thirteen used internally can be extracted if required.

Sharp PC-1211

This pocket computer has a flexible memory, with 1424 steps or 178 stores, interconvertible in an 8:1 ratio. The stores 1 to 26 are named A to Z and these letters must be used as the names for variables (e.g. using the name AA gives a syntax error signal). Entire phrases can be assigned to a key, e.g. Z could mean PRINT or, say, $X * X + EXP(X) - Y * Y$; this leads to quicker program writing. Strings up to seven characters long can be handled. One common error when writing an algebraic expression such as $x + 2y$ in BASIC is to write $X + 2Y$ instead of $X + 2 * Y$, leading to a syntax error and program halt. Remarkably, the PC-1211 correctly interprets the expression 2Y as $2 * Y$. Further, it can accept expressions as input e.g. $\sqrt{2} + 3 * B$ is an acceptable input, whereas it would produce an error signal on any other computer which I know. Calculated GOTO is possible e.g. GOTO 10*N. The 1424 steps on the PC-1211 represent somewhat more than 1.5K in usual microcomputer terminology.

Sinclair ZX-81

This small computer operates ideally in conjunction with a portable television with a continuous tuning control (I use mine with a Bush Ranger 3 and find it very easy to tune to the computer's frequency band in channel 36 UHF). The ZX-81 has a QWERTY typewriter keyboard, but each key also has multiple uses (e.g. LET, INPUT, RUN, GOSUB, COS are assigned to specific keys) so that the full alphabetical typing of control words is not necessary. The keys are actually 'touch' squares on a smooth panel. Array subscripts as in, say, M(I, J) can be 1, 2, 3, etc, not 0. The computer runs in two speed modes, controlled by the commands FAST and SLOW. Although only one statement per line is allowed, that statement may contain a long mathematical expression which spills over into the next *screen* line. (On most microcomputers the nominal 'line' can take up more than one display line.) The ZX-81 will generate random numbers for use in calculations using probabilistic models. It can accept simple expressions, e.g. $5 + (2 * 3)$ as input values for variables.

The CBM Pet

The common instruction PRINT is produced by a single ? key on the Pet keyboard. Arrays can be multiple e.g. a cubic array M(I, J, K) can be stored and used. Array subscripts can be 0, 1, 2 etc. Variables are initialised to the value zero on the RUN command and do not have to be declared at the start of a program. For example, the following statement (in which no LET is needed)

$$M = N + 23$$

would be executed even if M and N were making their very first appearance in the program. Stores would be allocated for M and N; with N initialised to zero, M would end up with the value 23. The Pet can generate random numbers between 0 and 1. It also has an internal clock, which is useful for doing speed tests on alternative algorithms (see §2.3). The timing of a calculation is performed as follows.

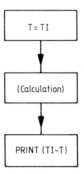

The variable T is set equal to the initial time, the calculation is performed, and the starting time T is subtracted from the present time to give the elapsed time in jiffies (1 jiffy $= 1/60$ s).

In a recent book M R Harrison [12] has pointed out that the ZX-81 can provide a timer if the television frame counter is used, together with the PEEK and POKE instructions (possessed by many microcomputers) which put numbers into or copy them out of specific locations. Thus the instruction POKE 16 436, 200 (for example) followed by POKE 16 437, 100 would set the counter to $200 + 256 \times 100 = 25\,800$ by setting the low and high bytes. Every fiftieth of a second the number is decremented by 1, so at a later time the quantity PEEK 16 436 + 256 ∗ PEEK 16 437 gives the current count. Subtracting this from 25 800 gives the elapsed time in fiftieths of a second. By using this approach I found that a delay loop (see solution 2.2) of form

```
10 FOR N = 1 TO Q
15 NEXT N
```

takes Q/50 s on the ZX-81 in SLOW running mode. In FAST mode the screen is turned off, so the frame counter remains fixed in value and does not give a timing indication. (There may still be a smart way to get at the computer's internal clock, but at the moment I don't know it.) The speed ratio between FAST and SLOW modes is roughly 4:1, and we can estimate relative running times for two programs by using the SLOW mode.

When a calculation is going to be run several times it is crucial to know what happens to the values of variables when the RUN command is given to start a new run. the Pet sets all variables equal to zero, and treats newly arising variables as discussed above. If we try the statement

```
LET M = N + 23
```

(with M and N making their first appearance) on a ZX-81 then it will stop, because it cannot find a value for N. However if an N value has been assigned earlier (e.g. LET N = 0) it will set up an M store and put M = 23 into it. Thus we can have new names on the left but not on the right of the =: we have to initialise each variable explicitly by some kind of statement. On the PC-1211 the M and N act rather like STO 10 and RCL 14 on a TI-58: (A to Z)≡(1 to 26). However the RUN command does not disturb the values of the variables, which will be whatever they were at the end of the last run, perhaps last week's run, since the PC-1211 has a permanent memory!

Exercises

1. The TI-58 calculator, when evaluating the square root of a real number x, gives $\sqrt{x}$ if $x > 0$ and a flashing display of $\sqrt{-x}$ if $x < 0$. How could this be useful in finding the roots of a quadratic equation with real coefficients?

2. If $\sqrt{2}$ is evaluated on a PC-1211 the result is 1.414 213 562. By supposing that $\sqrt{2}$ can be written as $1.414 + H$, derive a quadratic equation for H. Solving this equation using the standard formula would involve taking $\sqrt{2}$ again. Proceed instead by giving an equation for H which has H on both left- and right-hand sides, so that starting with the input $H = 0$ on the right we can use the equation iteratively to get a correct H value. Show that $\sqrt{2}$ can then be obtained to three more digits than the value quoted above.

Can you see how to convert this procedure into a general algorithm which would give high accuracy square roots for ten digit numbers between 1 and 10?

3. If the line

LET X = X/5 + 3

appears in a BASIC program it causes the microcomputer to take X from its location, work out the right hand side and then put back the result into the X location as the new X value. If the same line (without LET, of course) appeared in a piece of algebra we would take it to be an equation for X, and would conclude that X = 3.75. As an amusing exercise, see if you can figure out a simple way to get the computer to treat the line as an equation and give X the value 3.75.

4. Logical tests involving IF (e.g. IF A > 4 THEN GOTO 100) are useful for controlling the flow of a calculation. On the Pet the word GOTO can be omitted if desired. Most microcomputers accept multiple conditions such as

IF R > = 0 AND R < = 1 THEN GOTO 100

Clearly, to achieve a reasonable degree of transportability of a program from one BASIC machine to another it is necessary to write it to include LET, THEN GOTO, etc in full. Even though this makes it include words which are not essential on some machines, those machines will still accept the program with its redundant words. To be safe, then, we could take a 'lowest common denominator' type of approach. On a TI-58 calculator we can do many of the things which a BASIC machine can do, but to perform conditional jumps we use the t register. When some number R is calculated, then it is effectively in the display (since we think through the calculation as if we were doing it manually). If the t register contains 0 then the program steps

2nd $x \geqslant t$ A

will make the program jump to step A if R (i.e. the display number) is greater than or equal to the t number (in this case zero); otherwise the calculation simply continues. Can you write the BASIC double condition above so that it could translate into a single t register test on the TI-58?

5. As already noted the PC-1211 and ZX-81 will accept expressions as input values for variables. In the case of the PC-1211 the expression can involve functions and also the values of other variables; we cited the example $\sqrt{2} + 3 * B$. The keystroke R/S in a TI-58 program will halt the program for input. This returns the machine to manual mode, so that we can work

out the input expression before pressing R/S to continue the calculation. For example, if we know that the B value is in store 2 we can use the keystrokes

$$3 \times \text{RCL } 2 + 2\sqrt{} = \text{R/S}$$

On the Pet the statement

INPUT "A"; A

makes the screen display A? This needs a number as input; using an expression will give the response REDO FROM START. Concoct one way of getting an expression accepted as input, using the TI-58 example.

Solutions

1. When the calculator works out $\sqrt{b^2 - 4ac}$ it will give either a steady or a flashing display. This tells us quite visibly whether the roots are real or complex. The following program would do the calculation and has been chosen to show some TI-58 features.

Input Data. a in store 1, b in store 2, c in store 3.

Program

RCL2 $\div$ 2 $\div$ RCL1 $=$ STO4	(1)
RCL2 $x^2 - 4 \times$ RCL1 $\times$ RCL3 $=$	(2)
$\sqrt{} \div 2 \div$ RCL1 $=$ STO5 R/S	(3)

2nd Lbl A

RCL5 $-$ RCL4 $=$ R/S	(4)
$-2 \times$ RCL5 $=$ R/S	(5)

2nd Lbl B

RCL4 $\pm$ R/S	(6)

Line 1 puts $b/2a$ into store 4 and lines 2 and 3 work out $\sqrt{b^2 - 4ac}$, divide it by $2a$ and show it at R/S (run-stop). If the display is flashing then it is the imaginary part of the roots. Pressing key B shows the real part. If the display at 3 is steady, pressing A gives the first real root and further pressing R/S gives the second root.

2. Setting $\sqrt{2} = 1.414 + H$ and squaring both sides leads to the equation

$$H = 0.00\,604/\,(2.828 + H)$$

which yields, after three cycles, starting from $H = 0$ on the right, a stable H value. This gives us

$$\sqrt{2} = 1.414 + H = 1.414\,213\,562\,373\,1$$

To get the same kind of process to work for other numbers we can use the BASIC program following on the PC-1211.

```
10 INPUT X
20 A = INT (1000 * √X̄)
30 B = A/1000 : R = X − B * B : H = 0
35 FOR N = 1 TO 3
40 H = R/ (2 * B + H)
45 NEXT N
50 PRINT B : PRINT H
60 GOTO 10
```

Line 20 takes $\sqrt{X}$, multiplies it by 1000 and takes the integer part, giving the 'sawn-off' value analogous to the 1.414 of our example. Line 30 works out the number analogous to the 0.000 604. Lines 40 to 50 give a loop which applies the formula three times to get *H*. Line 50 shows the two portions of the high accuracy square root. Line 60 takes us back for the next X. (This program doesn't work perfectly on the Pet or the ZX-81, as I shall note in chapter 2, so must be slightly modified.)

3. If we get the computer to just keep on repeating the statement the number in the X location will move towards and finally reach 3.75. This is an example of a convergent iterative solution of a polynomial equation (see §3.2). If the starting value of X is 0 it takes 17 cycles on a PC-1211 for X to stabilise on the value 3.75. Although the method can work even with more complicated functions on the right it does not always work e.g. it will not work if we have $2 * X + 3$ on the right-hand side. Try it and see! To make the statement repeat many times one procedure is to put it in a loop like that used in exercise 2.

4. The function $R(R − 1)$ is negative only if $R > 0$ and $R < 1$.
    The following lines would suffice

    LET RR = R * (R − 1)
    IF RR < 0 THEN GOTO 100

(The name need not be RR, but should not have been used elsewhere in the program.)
The following steps on the TI-58 would do the trick

    STO 5 x^2 − RCL5 = ±
    2nd $x \geqslant t$ A

The R value is kept in store 5 and not lost. If we don't need it then we can use the algebraic result $x(x − 1) = (x − \frac{1}{2})^2 − \frac{1}{4}$ and replace the first line by

    $−0.5 = x^2 − 0.25 = \pm$

Think it through! There are other clever ways of doing double conditional jumps (e.g. using flags) on the TI-58 but the simple tricks described here correspond to straightforward BASIC statements.

5.   The statement STOP halts the program and transfers control to the operator,
     who can then input

$$A = 3 * B + SQR\,(2)\ \ RETURN$$

followed by CONT RETURN to continue the calculation. This procedure
is useful for forcing into the machine a manual change in the values of the
variables during the course of an interactive computation. In the above
example a PRINT "A=" statement before the STOP would simulate the
usual input statement and also remind the operator to start off with A=.

1.5   Flowcharts

When constructing an algorithm or a program most scientists use some sort of
flowchart to clarify the 'tactics' to be employed. The structure of the programs
in this book is fairly simple, but I do give flowcharts for some of them *after* the
program, with each box showing which program lines carry out the operation
mentioned. The square root program from solution 2 can be set out as follows.

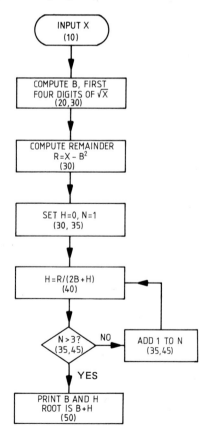

Notes

1. D Alcock 1977 *Illustrating Basic* (Cambridge: Cambridge University Press)
2. J P Killingback 1981 *The Creative Use of Calculators* (Harmondsworth: Penguin)
3. I Gladwell and D K Sayers (editors) 1980 *Computational Techniques for Ordinary Differential Equations* (London: Academic Press)
4. E T Bell 1953 *Men of Mathematics* (Harmondsworth: Penguin)
5. C Grey *Enterprise 2115* (London: Merit Books)
6. E U Condon and G H Shortley 1935 *The Theory of Atomic Spectra* (Cambridge: Cambridge University Press)
7. L Fox and D F Mayers 1968 *Computing Methods for Scientists and Engineers* (Oxford: Oxford University Press)
8. J Ciardi (editor) 1950 *Mid-Century American Poets* (New York: Twayne Publishers Inc)
9. J Maniotes, H B Higley and J N Haag 1971 *Beginning Fortran* (New York: Hayden Book Co Inc)
10. C Duff *The Basis and Essentials of French* (London: The Orthological Institute and Thomas Nelson & Sons Ltd)
11. E Hemingway 1952 *The Old Man and the Sea* (London: Jonathan Cape)
12. M R Harrison 1981 *Byteing Deeper into your ZX-81* (Wilmslow, Cheshire: Sigma Technical Press)

2 Tuning the instrument

2.1 General comments

If a scientist wishes to learn something about a situation, then he will use
a mixture of empirical and theoretical procedures to increase his knowledge.
In a very precise sense (which has relevance in atomic physics) we can say that
only knowledge about interactions is gained, since we must interact with a system
in order to learn about it. How far that knowledge can be translated into knowledge about the investigated system itself is still a matter of some debate e.g. in
connection with different interpretations of quantum mechanics. For a *mathematical* problem we are usually more confident that there does exist in principle
a solution, which we could approach by using a more and more accurate and
speedy computer in an appropriately designed program. (That 'accuracy' is
a software concept as well as a hardware one is, of course, part of my theme in
this book.) Real computers, however, give rounding errors, overflow problems,
etc so that we are always looking at our problem with some dirt on the telescope
lens, so to speak. One obvious point is that computers work internally with
binary numbers, but have input and output in decimal form for the operator's
convenience. The quirks of the translation process mean that a machine such
as a ZX-81 might appear to have 9 or 10 digit accuracy, depending on which
part of the real number region we are using. A necessary prelude to a serious
study of a numerical problem is the task of calibrating the apparatus, so that we
can be sure that later results refer to the mathematical problem and not to the
computer's own internal characteristics. In this chapter I outline a few typical
ways of discovering (and correcting) the weak spots in a microcomputer's
armour, and then discuss how the correct analysis of a problem can help to
improve speed and accuracy.

2.2 Significant Figures. Tests of Accuracy

Test 1
We can use the program

```
10 INPUT A
20 PRINT A
```

and input the number 1.234 567 8912 on run 1 and the number 1.234 567 8989 on run 2. The screen will take as long a number as we wish, but it will not all be accepted internally. Results for my four typical microcomputers are as follows:

	Run 1	Run 2
TI-58	1.234 567 891	1.234 567 898
PC-1211	1.234 567 891	1.234 567 898
ZX-81	1.234 5679	1.234 5679
Pet	1.234 567 89	1.234 5679

These indicate that the ZX-81 and the Pet round numbers scientifically while the PC-1211 truncates. The TI-58 numbers are those which can be keyed in until the display is full. However, the two test numbers can be forced in fully on the TI-58; we formally do the sum $1.234\,567\,891 + 2 \times 10^{-10}$. Indeed in this way we can force in a number as long as 1.234 567 891 222, exploiting the three guard digits carried in the TI-58.

Test 2
The program

```
 5 INPUT A
10 INPUT B
15 INPUT C
20 LET X = A + B + 0.01 − C
25 LET Y = A + B − C + 0.01
30 PRINT X, Y
35 GOTO 5
```

can be used to illustrate how non-commutativity of addition can occur on a computer. For example if we input $A = 10^K$, $B = 3 \times 10^K$, $C = 4 \times 10^K$ for positive integer K, then at some value K1 the X value should go off in one or more digits from the exact value 0.01, because the small number 0.01 will be masked by the enormous value of A + B. At a larger value K2 the X value will be zero, because 0.01 will go entirely unnoticed and C will cancel A + B. For my four typical machines I found that Y came out as 0.01 for any K value, whereas the X calculation gave the results

	K1	K2
TI-58	10	10
PC-1211	10	10
ZX-81	0	8
Pet	0	8

These peculiar results arise because the ZX-81 and the Pet do not subtract numbers perfectly. The error is very small, usually in the last one or two digits only, but has to be remembered if we need an exact subtraction (as for example in the special square root algorithm in exercise 2 of Chapter 1). One way to handle this is to convert the numbers being subtracted to integer form. For example the Pet gives the result 4.559 988 16E-05 if asked to work out $1.012\,3456 - 1.0123$, but if we input A and B and then use the statements (with LET omitted)

$$A = A * 1E7 : B = B * 1E7$$
$$D = (A - B)/1E7$$

then we get the correct result 4.56E-05. However, the 'clever' statement

$$D = (A * 1E7 - B * 1E7)/1E7$$

does not give the exact result, but rather 4.559 9942E-05.

The ZX-81 can be treated similarly but (as of September 1981) has a further eccentricity. If we work out $A \pm B$ with $A = 10^K$ and $B = 1, 2, 3$, etc, then for large K we expect to get the result 10^K because B will only affect digits beyond those which the machine is handling. The ZX-81 bears out this requirement for $A + B$, but for *all* small numbers B it gives the following values of $A - B$ when $K > 9$:

K	10	11	12
A − B	2.717 9869E10	2.374 3895E11	2.099 5116E12

To get round this we could use statements such as

LET D = A − B
IF D/A > 1 THEN LET D = A

i.e. we would have to teach the machine how to subtract (just as in machine code we have to teach a computer how to multiply by combining additions and shifts).

In general, addition on all the four machines which I have tested here is carried out without much trouble, but the Pet is the one showing the slight eccentricity. To do an integration numerically we have to increase the x coordinate in steps of size h, where h is the integration stripwidth. The simple statements

```
 5 INPUT H
10 INPUT X
15 LET X = X + H
20 PRINT X
25 GOTO 15
```

should suffice to move along the x axis. My personal preference is to use very simple h values, 0.01, 0.02, etc, to get rid of any surplus rounding error effects which might arise if I use 'clever' values involving many significant digits. I found that the TI-58, the ZX-81 and the PC-1211 all count perfectly from $X = 0$ if $H = 0.01$ is used in the above program. The Pet, however, goes off slightly above $X = 0.79$ and wanders around the exact value, with the last one or two digits (in ten) being wrong. The use of integer arithmetic helps to remove this problem (except for one or two particular X values). Thus, when integrating on the Pet I use the following replacement for statement 15, if X starts at 0:

```
15 N = N + 1 : X = N * H
```

(Remember that we need no LET on the Pet and can use several statements per line).

Test 3
This one is a rather cruel one which *must* beat every computer! According to pure mathematics the number e can be defined as

$$e = 2.718\,281\,828\ldots = \underset{N \to \infty}{\mathrm{Lt}}\ (1 + N^{-1})^N.$$

The following BASIC program gets the computer to work out the quantity $E(N) = (1 + N^{-1})^N$ for $N = 1, 2, 4, 8$ etc. (Trace it through for yourself: LETs are omitted and multiple statement lines are used for brevity.)

```
10 N = 1 : C = 0
20 N = 2 * N : C = C + 1
30 R = (N + 1)/N : E = R
40 FOR M = 1 TO C
50 E = E * E : NEXT M
60 PRINT E, N : GOTO 20
```

Pure mathematics shows that $E(N)$ increases with N until it reaches the limiting plateau value e. All four machines initially give increasing $E(N)$ but eventually give the repetitive output 1, since the N^{-1} becomes 'invisible' relative to 1. The Pet gives a plateau at 2.718 281 38 and the ZX-81 gives one at 2.718 2814, followed by a decline to 1. The PC-1211 E(N) oscillates above and below e before falling to 1. The TI-58 gives a plateau at 2.718 2775. In constructing the e test I did not use the powering function ($\uparrow$, $**$, $\wedge$ or y^x) since this uses the exponential function, involving e. Using the instruction E = R $\uparrow$ N on the Pet, for example, gives an E(N) which increases right through the true e value up to 128 before suddenly dropping to 1.

2.3   Some Speed Tests

The following tests illustrate how the detailed way in which a program is written can have a marked effect on the running speed. I start with the program (A)

```
10 FOR N = 1 TO 100
20 LET A = 5.123 45 * (N ↑ 3)
30 NEXT N
40 PRINT "END"
```

and vary it by (B) using N * N * N in line 20 to replace N $\uparrow$ 3, and then also (C) using C in place of 5.123 45, with a new statement before the loop setting C equal to 5.123 45. Running times in seconds for my test machines are shown below: for the BASIC machines they show that it is usually of benefit to use the cumbersome 'repeated product' form for integer powers and to use dummy variables to replace constants.

	A	B	C
TI-58	90	93	95
PC-1211	100	50	47
ZX-81	12	1.4	1.4
Pet	7.9	3.0	1.0

The powering operation ($\uparrow$, $**$, $\wedge$ or y^x) of course has to be used for general non-integer powers of a number, but on the ZX-81 and TI-58 it will not give correctly the powers of negative numbers, even when these are well defined real numbers e.g. $(-2)^3 = -8$. The prescription N * N * N to get N^3 works for either sign of N. The trick used in (C) above of using a dummy variable C is even more valuable if we have an expression instead of 5.123 45 (e.g. 37.2 + 4 * SQR(L), where L is fixed and independent of the loop variable). Instead of working out the expression 100 times we would work it out once before the loops begin and call it in as the dummy variable C on each cycle.

2.4   Subroutines

The statement GOSUB 500 will make the program go to line 500 and then execute statements down to the statement RETURN which means 'GOTO the statement after the GOSUB which brought you here.' Subroutines usually contain some set of statements which have to be used many times during a program and avoid the need to write that set explicitly many times in the program. Subroutines would be needed on grounds of economy even if computers were of infinite memory. Once a subroutine for, say, multiplying two complex numbers has been perfected then it can be used as a standard component in any long program which needs such an operation. There is also another use for subroutines: the correction of defects in the computer! I list a few examples.

Integer Arithmetic
If we want to do highly accurate subtractions several times during a calculation then a subroutine which transforms to integer subtraction as explained in §2.2 (Test 2) could be used. Note that a temporary change of name is often necessary for the variables. Thus, we may wish to know $Z = X - Y$ in the main program, but the subroutine statements may be written to find $D = A - B$. To take care of this we use the statements

$A = X : B = Y$
GOSUB 500
$Z = D$

which leave X and Y unaffected and make Z equal $X - Y$. It is common for beginners to forget to 'line up' the variables between the main program and the subroutines, or to let a program 'fall into' a subroutine by forgetting that the main program will eventually get down to line 500 and execute it unless stopped by an END, GOTO, etc.

Overflow Suppression
If one of the numbers in a calculation exceeds the overflow value then the program will halt with an overflow error indication. It is useful to have a trick which avoids a program halt when some variable's *absolute* size exceeds overflow. For many calculations, particularly quantum mechanical ones which embody some kind of linear operator mathematics, it may be only the *relative* sizes of some finite number of variables which matters. Suppose, for example, that these variables form a linear array $A(1)$ to $A(8)$ and that they are known to be of roughly the same order of magnitude throughout the calculation. With an overflow at IE38 we could play safe by using statements such as

```
120  IF A(8) > IE35 THEN GOSUB 500
500  FOR N = 1 TO 8
510  LET A(N) = A(N)/1E6
520  RETURN
```

Of course, whether a scaling down factor of a million is the appropriate one would depend on our experience of the particular calculation. Overflow suppression is only rarely needed for the TI-58 and the PC-1211, which can handle numbers up to 1E100.

Array Shuffling

In some calculations based on recurrence relations we may have a statement such as

$$50 \quad T(N+2) = A * T(N+1) + B * T(N) + C$$

or one involving more than three T values. If T has been declared as an array there will be some maximum possible N value determined by the computer's RAM capacity. To complete the calculation we may have to go to greater N values; for example to sum a series such as

$$S = \sum_{1}^{\infty} T(N)$$

we have to keep going until S converges (§7.4 provides an example). 'Array shuffling', as I call it, uses three stores, since we only need to keep three $T(N)$ at a time to do the calculation and so only need to know $T(0)$ to $T(2)$ and S. To 'shuffle along' the array elements we could use a subroutine with statements such as

```
500  FOR N = 0 TO 1
510  T(N) = T(N + 1)
```

and replace T(N) by T(0), T(N + 1) by T(1) and so on in the original statement embodying the recurrence relation:

$$50 \quad T(2) = A * T(1) + B * T(0) + C$$

On a ZX-81, which doesn't allow subscript 0 for an array element, we would add one to the subscripts, but would have to remember this carefully if the coefficients A, B and C depend on N (as they do for the calculation of §7.4).

Mathematical Subroutines

If the computer does not have natural BASIC instructions for operations such as adding and multiplying complex numbers and matrices then special subroutines to do the job will be required. Once written they can be used over and over again in many programs. Alcock [notes 1.1] quotes some versions of BASIC which have command words to multiply and invert matrices but I have not personally used a microcomputer with such built-in facilities.

2.5   Labour-saving analysis

§2.2 and 2.3 gave a few simple ways of getting extra speed and accuracy from a BASIC program. I now want to give two examples of using analysis to cut down the number of calculations which have to be performed. First, consider the family of integrals

$$I_N(\lambda) = \int_0^\infty x^N \exp(-\lambda x^4)\, dx \tag{1}$$

with N a positive integer or zero and λ a positive real number. By differentiating with respect to λ we find

$$I_{N+4}(\lambda) = -\frac{\partial}{\partial\lambda} I_N(\lambda). \tag{2}$$

However, by changing to the variable $y = \lambda^{1/4} x$ we also find

$$I_N(\lambda) = (\lambda^{-1/4})^{N+1} I_N(1). \tag{3}$$

Combining the last two results gives

$$I_{N+4}(1) = \left(\frac{N+1}{4}\right) I_N(1). \tag{4}$$

Thus, to get I for any allowed N and λ we only need to calculate explicitly the integrals $I_0(1)$ to $I_3(1)$. This example has many similarities to those arising in the theory of hypervirial relations (§9.6) except that in the latter case the integrals are interpreted as quantum mechanical expectation values.

My second example is a simple Schrödinger equation which has to be treated numerically;

$$\frac{-\hbar^2}{2m} D^2 \psi + \lambda x^4 \psi = E\psi. \tag{5}$$

Although $\hbar$, m, e, etc appear in the standard textbook equations of quantum mechanics, nobody in his right mind wants rotten numbers such as 6.6252×10^{-27} on the loose in his computer program! In atomic theory computations, for example, special atomic units are employed in which the ground state energy of the hydrogen atom is $-\frac{1}{2}$, whereas in 'ordinary' units it is $-\frac{1}{2}m^2e^4\hbar^{-2}$. For the one dimensional example above we try changing the length coordinate to $y = Kx$. In y language the equation becomes

$$\frac{-K^2\hbar^2}{2m} D^2\psi + \lambda K^{-4}y^4\psi = E\psi. \tag{6}$$

Now we choose K so that $K^6 = 2mh^{-2}\lambda$ and multiply the equation through by a factor $K^4\lambda^{-1}$. We get

$$-D^2\psi + y^4\psi = K^4E\lambda^{-1}\psi. \tag{7}$$

If the bound state boundary condition is $\psi \to 0$ as $x \to \pm\infty$, then it takes the same form in y language. We can do the entire calculation for the Schrödinger equation (7), with a simple left-hand side for numerical work, and simply divide by the factor $(K^4\lambda^{-1})$ to get the energy levels for the original Schrödinger equation. Since we have

$$\lambda K^{-4} = \lambda^{1/3}(\hbar^2/2m)^{2/3} \tag{8}$$

it follows that the energy eigenvalues vary as $\lambda^{1/3}$ with λ. In this case we could have anticipated the result from dimensional analysis, by taking $(\hbar^2/2m)$ to be of type $[E][L]^2$ and λ to be of type $[E][L]^{-4}$ with $[E]$ = energy and $[L]$ = length.

Throughout this book I try to emphasise the value of analysis in computing: in particular I illustrate how one can sometimes show that some quantity A, which is difficult to compute, is equal to some other quantity B which can be obtained with less difficulty. For example, to work out the kinetic energy expectation value we should formally integrate the product $-\psi D^2\psi$; this involves a second derivative, which may be messy if ψ is complicated. However, a little textbook integration by parts shows that the integral of $(D\psi)^2$ will give the same result, while needing only the less complicated first derivative. In Appendix 1 I show that it is even possible in principle to do the calculation without derivatives at all. Of course, a computer cannot differentiate even once: it has to be instructed to simulate differentiation by using finite-difference calculations, unless we explicitly give it $D\psi$ as a function (i.e. we differentiate ψ analytically before writing the program). In §9.4 I show how to evaluate expectation values from energy calculations without even knowing the wavefunction! This is just about as far as it is possible to go, and illustrates dramatically how a careful use of analysis can help in numerical work.

Exercises

1. Work out on some microcomputer the fraction $(x^2 - 1)/(x - 1)$ and compare it with $x + 1$, which it should equal according to pure algebra. Try the x values 1.1, 1.01, 1.001 and so on and explain your results.

2. Suppose that we want to introduce a time delay into a program e.g. a game program where we want to slow down the moves. Look at the speed test programs of §2.3 and see if you can think of a way to produce a delay.
3. Show how to modify the high accuracy square root program (exercise 1.2) for a computer which doesn't subtract perfectly, in order to get square roots for eight digit numbers between 1 and 100.
4. Consider a linear relation of the form

$$ax + by + cz + d = 0 \qquad (9)$$

with a, b, c, d known. If we wish to find x, say, knowing y and z, we can proceed by using the program line

LET X = (B * Y + C * Z + D)/(−A)

but would need a *different* line to find y from a known (x, z) pair. Try to construct a procedure which uses as its main component the subroutine

500 LET F = A * X + B * Y + C * Z + D
510 RETURN

which just evaluates the entire sum of four terms.

5. Construct a subroutine to work out the quantity $(X1 + i\,Y1)(X2 + i\,Y2)^K$ with $K = \pm 1$. The input from the main program is to be the real parts $(X1, X2)$ and imaginary parts $(Y1, Y2)$ of two complex numbers, together with a K value (1 for multiply, −1 for divide).
6. The *nested multiplication* procedure for working out the value of a polynomial proceeds as outlined below.

If $P_N = \Sigma_0^N A(n)z^n$, set $Q(N) = A(N)$ and use the following recurrence relation (for decreasing n)

$$Q(n) = A(n) + zQ(n + 1). \qquad (10)$$

The last value obtainable is $Q(0)$ which equals the value of P_N. For the case $N = 2$, for example,

$$2z^2 + 3z + 1 = (2z + 3)z + 1. \qquad (11)$$

Nested multiplication is compact in that it uses few multiplications. Consider the problem of working out the sum to N terms, then $N + 1$ terms, and so on, for an infinite power series. Would nested multiplication have any drawbacks because of the way it works down from N to 0? Is it possible to go forwards from 0 to N without explicitly working out each term $A(n)z^n$ separately?

7. Consider the one-dimensional Schrödinger equation

$$-\alpha D^2 \psi + \mu x^2 \psi + \lambda x^4 \psi = E\psi \qquad (12)$$

which involves three real parameters. To find the energy $E(\alpha, \mu, \lambda)$ of some particular bound state as a function of α, μ and λ is apparently a lengthy task. Show by using the approach of §2.5 that we can reduce the problem to a one parameter problem.

8. Show by partial integration that the integral

$$F(N) = \int_0^\infty t^N e^{-t}\, dt \tag{13}$$

has the value $N!$ (factorial N) when N is a positive integer. For non-integer N the integral is still calculable numerically and gives the factorial function. Show by a change of variable that the integral I_N of §2.5 can be expressed in terms of the factorial function.

Solutions

1. At $x = 1 + 10^{-5}$ the Pet and ZX-81 give the value of the fraction as exactly 2. If we write x^2 as $(1 + h)^2 = 1 + 2h + h^2$ then it is clear that h^2 is 10^{-10} if h is 10^{-5}; the h^2 term thus becomes invisible if we don't have eleven signifi-cant figures, since there is a leading 1. In fact, because of imperfect subtraction involving the $2h$ term the Pet produces the result 1.5 at $x = 1 + 10^{-7}$ and below.

2. To get a time delay we use an empty loop, which we can calibrate using a stop watch, or an internal timer on the Pet or ZX-81. I tried the following program on the Pet;

```
 5 INPUT Q
10 T = TI
20 FOR N = 1 TO Q : NEXT N
30 PRINT (TI-T) : GOTO 5
```

The choices $Q = 1000$ and 2000 gave time values of 85 and 170 jiffies, so that on my Pet a delay of 1 second goes with $Q = 700$, for this particular layout of the statements. (See §1.4 for the ZX-81 timer.)

3. We can use the program of solution 1.2 with the modification

```
30  B = A/1E3 : Y = X * 1E7 : Z = B * B * 1E7
32  R = INT (Y − Z + 0.5)/1E7
```

Line 30 introduces integer arithmetic and the extra 0.5 in line 32 allows for the possibility that even the integer subtraction might be slightly wrong! The 10^7 factor is right to ensure that an eight digit number between 1 and $9.9\dot{9}$ is converted to an integer.

4. To find x we use the statements

$$X = 0 : GOSUB\ 500 : X = -F/A$$

with analogous lines to find y or z.

5. My version, which will probably differ from yours, is

```
500  DF = 1 + K + (1 − K) * (X2 * X2 + Y2 * Y2)
510  MF = 2/DF
520  RR = (X1 * X2 − K * Y1 * Y2) * MF
530  IR = (Y1 * X2 + K * X1 * Y2) * MF
540  RETURN
```

The subroutine returns the real and imaginary parts (RR and IR) of the result. The names used are arbitrary, but must not 'interfere' with the variables in the main program. To add the complex number X3 + i Y3 to the product we simply add +X3 in line 520 and +Y3 in line 530. We can then use the subroutine in a nested multiplication evaluation (exercise 6) of a complex polynomial. This would be useful in applying Newton's method (§3.3) to find a complex root of a polynomial equation.

6. To work out the sum to $N+1$ terms of a power series would involve working down afresh from $N+1$ to 0, without using our knowledge of the sum to N terms. However, we can write P_N as follows

$$P_N = z^N \sum_0^N A(N-n) y^n \tag{14}$$

with $y = z^{-1}$. This reverses the order of the coefficients, and leads to the program:

```
5   INPUT Z : INPUT N : DIM A(N)
10  Q = A(0) : F = 1
20  FOR M = 1 TO N
30  Q = A(M) + Q/Z
40  F = Z * F
50  P = Q * F : PRINT P, M
60  NEXT M
```

This is my solution, but it is not the only one, I am sure. I have avoided using arrays except for the coefficients A(M), but even that can be avoided if we use the following line 30

```
30  INPUT A : Q = A + Q/Z
```

Alternatively, we can make everything an array and store all the results of the calculation in the microcomputer! I have used the program above to sum the series of terms

$$S_N(z) = \sum_1^N n^{-4} z^n \qquad (z = 1). \tag{15}$$

At $N \sim 270$ the sum stabilises to $1.082\,323\,22$ on the Pet and to $1.082\,323\,208$ on the PC-1211. The correct value is $\pi^4/90$, which equals $1.082\,323\,234$. The error is probably due to the evaluation of n^{-4} for large n values, since a direct addition of the terms gives the same result on the Pet.

7. Introduce the variable $y = Kx$; multiply the resulting equation through by K^6 and then divide it by μK^2. Choose K so that $\alpha K^4 = \mu$. The resulting equation is

$$[-D^2 + x^2 + \lambda(\alpha\mu^{-3})^{1/2} x^4]\psi = (\mu\alpha)^{-1/2} E\psi \tag{16}$$

if we use the symbolic name x again for the independent coordinate. This result shows that

$$E(\alpha, \mu, \lambda) = (\alpha\mu)^{1/2} E(1, 1, \lambda') \tag{17}$$

where $\lambda' = (\alpha\mu^{-3})^{1/2}\lambda$. The values of α and λ are usually taken to be positive to ensure the existence of well defined bound states, but if $\lambda > 0$ then μ can be positive or negative and still give bound states. (The x^4 term dominates at large distances and keeps the particle from escaping.) A negative μ inside the square root signs causes trouble, but re-tracing the calculation shows that we should use $|\mu|$ instead of μ and have $-x^2$ in the final equation. We thus have to solve separate problems for $\mu > 0$ and $\mu < 0$; this particular problem still arouses some interest in the research literature and is related to the so-called soliton phenomena of field theory. In our example the ground state ψ is centred on $x = 0$ for $\mu > 0$, but for $\mu \ll 0$ it will be concentrated around the potential minimum at $x = d$, where $2\lambda d^2 = |\mu|$. It is perhaps of interest to note that dimensional analysis will work for this problem. If we use only energy and length, [E] and [L] and look for dimensionless combinations of α, μ, λ and E, we can set $\alpha \equiv [E][L]^2$, $\mu \equiv [E][L]^{-2}$ and $\lambda \equiv [E][L]^{-4}$. Using only α, μ and λ we find $\alpha\mu^{-3}\lambda^2$ as a dimensionless combination. $E(\alpha\mu)^{-1/2}$ is another and so we can surmise that the following relationship holds,

$$E(\alpha, \mu, \lambda) = (\alpha\mu)^{1/2} f(\alpha\mu^{-3}\lambda^2) \tag{18}$$

where f is an unknown function. (The reader disturbed by the sleight of hand here may note that if we go down to [M], [L], [T] dimensions we have three of them; from four variables this gives at most two dimensionless quantities.) From the relationship above we can see that

$$E(\alpha, \mu, \lambda) = (\alpha\mu)^{1/2} E(1, 1, \lambda') \tag{19}$$

where $\lambda'^2 = \lambda^2 \alpha\mu^{-3}$. Taking $|\mu|$ instead of μ doesn't affect the dimensional analysis, so the $\mu < 0$ case is implicitly contained in the reasoning.

8. Integrating by parts gives the result $F(N) = NF(N-1)$. This is sufficient to show that $F(N) = N!$ if we note that $F(0) = 1$. The change of variable $y = x^4$ takes the integral

$$I_N = \int_0^\infty x^N e^{-x^4} dx \tag{20}$$

into

$$I_N = \tfrac{1}{4} \int_0^\infty y^K e^{-y} dy = \tfrac{1}{4} K! \tag{21}$$

where $K = \tfrac{1}{4}(N-3)$. The relationship between I_{N+4} and I_N, which we can write as $4I_{N+4} = (N+1)I_N$, follows from the fact that adding 4 to N also adds 1 to K, while $(K+1)!$ is simply related to $K!$ Basis functions such as $x^N e^{-\beta x}$ or $x^N e^{-\beta x^2}$ are often used in quantum mechanics, and so integrals involving them often reduce to factorial function expressions.

3 The iterative approach

3.1 Introduction

In the previous chapters I have emphasised my view that iterative or recursive techniques are well suited for calculations on a microcomputer with a limited RAM capacity. In this chapter I give several specific examples of how to set up an iterative method for handling problems. One common way to refer to successive estimates in an iterative process is to use the symbols x_n and x_{n+1}, whereas I use x and y. My input-output point of view (§3.2) seems appropriate for computer work and the simple standard notation dy/dx or y' can be used in discussions of convergence properties. I describe the simple theory in §3.2 and give Newton's method, with a specific cubic equation example, in §3.3. An interesting example of a conflict between universality and accuracy for a program arises in that discussion. The iterative procedure for calculating inverses, described in §3.4, is one of the most useful simple methods in quantum mechanics, and has links with matrix theory, perturbation theory (§9.3), Padé approximant theory (§6.3) and operator resolvent theory (§12.6). In §§3.5 to 3.7 I outline how some matrix problems can be handled by an iterative approach. In particular I indicate how the Gauss–Seidel method can be rendered applicable even for matrices which are not diagonally dominant. The matrix folding method of §3.6 is really an application of Brillouin–Wigner perturbation theory to a numerical problem, and involves one of the nicest new programs which I have devised while writing this book. Exercise 4 introduces the Aitken process for treating sequences and series; this process plays a role in chapters 5 and 6.

3.2 The Input-Output Approach

Consider the following equations, with β an arbitrary number;

$$f(x) = 0 : y = x + \beta f(x). \tag{1}$$

If we regard the second equation as a computational prescription (i.e. input x, output y) then clearly with perfect arithmetic we only get output = input if the input x is a root of the first equation. It is easy to construct a loop program which keeps on working out y and putting it back as the input x for the next cycle. If the procedure converges to give a steady y value then we have a root of the equation $f(x) = 0$. However, the procedure might not converge, either to the root which we want or to *any* root. The parameter β can be adjusted to change the convergence properties of the procedure. Suppose that the input x is equal to $r + h$, where r is a root of the equation and h is the error, which we provisionally regard as 'small' in the sense of the calculus. The output y will be as follows, remembering that $f(r) = 0$;

$$y = (r + h) + \beta f(r + h)$$

$$= r + h + \beta f(r) + \beta h f'(r) + \ldots$$

$$= r + h \ \ [1 + \beta f'(r)] + \ldots. \tag{2}$$

If h is sufficiently small for the higher terms to be neglected and if $|1 + \beta f'(r)| < 1$ then the error is reduced on each cycle and the process will converge to the root r. This analysis is clearly a *local* one in that we take h to be small when we start, whereas it may not be so if we just put in an arbitrary initial guess x_0. To improve the analysis we can use the Taylor series with remainder, as studied in traditional calculus. In this case it simply amounts to saying that $[f(r + h) - f(r)]$ must equal h times the slope of the f curve at some point between r and $r + h$. (Draw a curve and see it!) This idea then gives us a result with no leftover terms;

$$y = r + h[1 + \beta f'(r + \theta h)] \tag{3}$$

where $0 \leqslant \theta \leqslant 1$. In the case that the quantity in square brackets has an absolute value less than 1 for all x between r and $r + h$ then the process converges even if h is initially large. In the case of a quadratic equation with no real roots, e.g. $f(x) = x^2 + x + 1$, making x_0 and β real locks the process on to the real axis, so that we cannot possibly arrive at a root. β or x_0 or both must be chosen to be complex if we want to get at the complex roots. (This would then need complex arithmetic, which could be handled by a subroutine such as that of solution 2.5.)

3.3 Newton's Method

Using the equations above we can perform a Taylor series expansion of the multiplying factor in the square bracket. We use the first version, i.e. $1 + \beta f'(r)$, but set $r = x - h$. This gives

$$1 + \beta f'(r) = 1 + \beta f'(x) - \beta h f''(x) + \ldots \tag{4}$$

and
$$y = r + h[1 + \beta f'(x)] + 0(h^2). \tag{5}$$

Making the choice $\beta = -1/f'(x)$ when the input is x will make the multiplying factor zero and give an error of order h^2 in y. This then yields what is usually called the Newton-Raphson method (or just Newton's method). The input-output formula for the method is

$$y = x - f(x)/f'(x). \tag{6}$$

The reason for the derivative juggling in the preceding analysis is that, although we have as our key fact that $f(r) = 0$, we do not *know* r and must produce formulae which mention only x, our current estimate of the root. If the desired roots are complex numbers then we have to use a complex initial estimate x_0. Longman [1] has given examples of the use of this method for finding complex roots. If instead of 'tuning' the β coefficient to the current x value we try some constant β value, then the error in y is of order h, not of order h^2 (i.e. we get a *first-order* iterative process, as opposed to a *second-order* one).

As an example of an integrated piece of analysis and computing I will now look at the problem of finding the roots of a cubic equation with real coefficients. Such equations appear in various branches of physics: in describing the amplitude jump phenomenon when an anharmonic oscillator is driven by a slowly varying frequency [2]: in calculating the depth to which a ball of density $\rho < 1$ will sink in water [3]: in calculating the volume V at given pressure and temperature for a van der Waals gas [4]. On looking at the equation in the form

$$Ax^3 + Bx^2 + Cx + D = f(x) = 0 \tag{7}$$

we see that we can always arrange for A to be positive, in which case f tends to $\pm\infty$ for $x \to \pm\infty$. Thus there must be at least one real root between $\pm\infty$, although there might be three. If some complex number z is a root then the complex conjugate z^* must be a root also, as we can see by taking the complex conjugate of the equation (with the proviso that the coefficients are real). Since complex roots occur in pairs, we must have either three real roots or one real and two complex roots. If we write the equation in the alternative form

$$A(x - r_1)(x - r_2)(x - r_3) = 0 \tag{8}$$

to display the roots r_1, r_2, r_3, then we can expand the triple product and compare the results with the original form of $f(x)$. We find

$$r_1 + r_2 + r_3 = -B/A : r_1 r_2 r_3 = -D/A. \tag{9}$$

If we have found one real root (r_1) then we can write the other two roots as $R \pm I$. Here R is simply the average of the two roots and is given by

$$R = -(r_1 + B/A)/2 \tag{10}$$

if we use the equation for the sum of the roots. Since $r_2 r_3 = R^2 - I^2$, the equation for the product of the roots gives us

$$I^2 = R^2 + D/(Ar_1). \qquad (11)$$

If I^2 is negative then the roots are $R \pm i|I|$; if it is positive they are $R \pm |I|$. We thus have an entirely real variable calculation which can give us the complex roots when they occur. The only thing needed to complete the job is to find the real root r_1 which we know *must* exist. A real x version of the Newton–Raphson method will suffice for this.

To apply the method we apparently need to work out both $f(x)$ and $f'(x)$. For the cubic equation we could explicitly tell the computer in its program that f' is $3Ax^2 + 2Bx + C$, but for a more complicated $f(x)$ it would be troublesome to state f' explicitly. What we can do is to replace $f'(x)$ by some finite-difference version which only needs evaluations of f. For example, if we take the rough approximation $h^{-1}[f(x+h)-f(x)]$ to $f'(x)$, with h small (say 0.01) the formula (6) becomes

$$y = x - hf(x)/[f(x+h)-f(x)]. \qquad (12)$$

We write one subroutine to evaluate f and jump to it whenever f is needed. To get a better estimate of $f'(x)$ we could use $(2h)^{-1}[f(x+h)-f(x-h)]$, but would have to make three f evaluations per cycle instead of two. In my experience this modification is not worthwhile. I did try a 'clever' version needing only two evaluations;

$$y = x - h[f(x+h)+f(x-h)]/[f(x+h)-f(x-h)] \qquad (13)$$

but it converges to roots which are slightly wrong. Why? Because if $f(x) = 0$ the numerator in the fraction is not exactly zero and we don't get $y = x$ as we do with the original simple form. This example is fascinating, because there is no doubt from numerical tests that the fraction in the second (wrong) version gives a more accurate value of $f(x)/f'(x)$ than the fraction in the first (successful) version!

A possible BASIC program to find the roots is as follows:

```
 5 INPUT A, B, C, D
10 INPUT X : H = 0.001
20 Y = X : GOSUB 80
30 G = F : X = X + H : GOSUB 80
40 X = Y − H * G/(F − G)
50 K = (X − Y)/X
60 IF ABS (K) > 1E − 7 THEN 20
70 GOTO 100
80 F = ((A * X + B) * X + C) * X + D
90 RETURN
100 R = −(X + B/A)/2
110 I = R * R + D/(A * X)
120 IF I < 0 THEN 150
```

130 Y = R + SQR (I) : Z = R − SQR (I)
140 PRINT X : PRINT Y : PRINT Z : END
150 I = SQR (−I)
160 PRINT X : PRINT R, I.

For particular machines slight variations will be needed, of course. The following comments explain some features of the program.

1. If the function F in line 80 is modified the program down to line 90 can be used to search for real roots of any equation $f(x) = 0$.
2. Note how the variables are properly named going into and coming out of the subroutine and how Y and G are used to store the preceding values of X and F.
3. The automatic convergence test in lines 50 and 60 can be dispensed with if the operator wants to use a PRINT statement and stop the program manually.
4. Lines 140 and 160 are suitable for a machine which can print two numbers on a line. The roots R ± iI give R and I on one line.

The following flowchart describes the layout of the calculation.

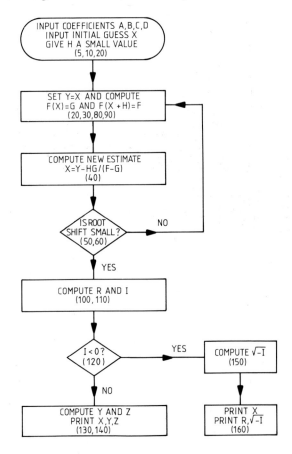

For the coefficient set $(4, 5, 6, 7)$ the program gave the roots $-1.207\,751\,19$, $-0.021\,124\,407 \pm i(1.203\,547\,97)$. For the coefficient set $(4, 5, -6, -7)$ it gave the roots $1.203\,768\,23$, -1 and $-1.453\,768\,23$. The sums of these roots one at a time, two at a time and three at a time agree exactly with the ratios $-B/A$, C/A and $-D/A$, showing that the roots are correct. For multiple real roots this simple program is not quite so good; it gives the roots $-0.999\,45$, $-1.000\,27 \pm i(0.000\,48)$ for the case $f(x) = (x + 1)^3$ on the Pet, with $x_0 = 2$. However, the results vary a little with x_0! This strange behaviour can be explained and remedied after a little analysis (exercise 4), but requires the operator to exercise his judgment when he sees this exceptional case arising. When the equation has three distinct real roots which one is found first depends on the initial x_0 value, but then the other two are produced automatically.

The program described above was intended to avoid the explicit statement of the derivative function $f'(x)$, and so parts of it can be used for applying Newton's method to equations of type $f(x) = 0$ for arbitrary f. If only polynomial equations are to be treated then there are several quick ways of working out f and f'. The following piece of program is one possibility, for the case $f(x) = A_N x^N + \ldots + A_0$. It is constructed to cut down on the number of 'look-ups' for subscripted array variables. Think it through!

```
50  F = A(N) : G = N * F
60  FOR M = N TO 1 STEP −1
70  A = A (M − 1)
80  F = F * X + A : G = G * X + (M − 1) * A
90  NEXT M : X = X − F * X/G
100 PRINT X : GOTO 50
```

3.4 Iterative calculation of inverses

The Newton–Raphson method, when applied to the problem of finding the inverse M^{-1}, gives the input-output prescription

$$y = x(2 - Mx). \tag{14}$$

The reader may verify that $x = M^{-1} + h$ yields $y = M^{-1} - hMh$. The form hMh (rather than Mh^2) was used here to highlight the valuable property that the results are valid even if M and x are *matrices* and not just real or complex numbers. Normally we would think of finding M^{-1} only for a square matrix M, but I discovered in the literature an interesting mathematical story which has only reached a conclusion comparatively recently [5]. If the set of linear equations $Mx = y$ (in brief matrix notation) has an M which is, say a 4×4 matrix, then the solution $x = M^{-1}y$ requires us to find M^{-1}, which is also a 4×4 matrix. However, if M is, say, a 3×5 matrix, so that we have five equations in three unknowns, then the iterative prescription given above can *still* be used. We find an M^{-1} which is a *generalised inverse* of M. M^{-1} is a 5×3 matrix and the product $M^{-1}y$ gives a 3-column x which is the *least squares solution* of the original set of equations. If the equations are written as $Mx - y = R$, where R is the so-called *residual* 5-column, $M^{-1}y$ gives the set of three x values which minimise the sum of the squares of the five residual elements. When M is square R can be reduced to zero by taking M^{-1} to be the traditional matrix inverse. For our specimen 3×5 case R cannot be made zero, but using the generalised inverse M^{-1} leads to a 3-column x which provides the best least squares fit to the set of five linear equations.

To apply the iterative prescription it is necessary to clarify the nature of each quantity in the formula. The formula takes the form

$$Y(A \times B) = X(A \times B)[2\mathbf{1}(B \times B) - M(B \times A) X(A \times B)] \tag{15}$$

where each matrix type is indicated and $\mathbf{1}$ is the unit $B \times B$ matrix. If M is a number then the iterative process converges if the initial X obeys the inequality $0 < X_0 < 2M^{-1}$. A theorem of comparatively recent origin [5] states that for the matrix problem the process converges if the initial matrix X is equal to $kM^\dagger$, where $(M^\dagger)_{jk} = (M_{kj})^*$ and k must be less than $2/|\lambda|$, where λ is the eigenvalue of largest modulus of $MM^\dagger$. A BASIC program which implements the iterative procedure for the generalised inverse was given by A Mackay [6]. The iterative process for reciprocals also is related to the Hylleraas principle of perturbation theory (§9.3) and the theory of Padé approximants (§6.3); indeed it has a fair claim to be one of the most useful iterative formulae in quantum mechanics!

3.5   The Gauss–Seidel method

When solving the set of linear equations $Mx = y$ for square M it is possible to get the solution for any particular given y *without* first finding M^{-1}. The Gauss–Seidel method is a simple iterative method which does this; I explain it by means of an example. Suppose that we have a set of three equations in the form $Mx = y$, with

$$M = \begin{pmatrix} 4 & 1 & 2 \\ 1 & 3 & 1 \\ 1 & 0 & 2 \end{pmatrix}. \tag{16}$$

The equation set $Mx = y$ can be rearranged to give

$$4x_1 = y_1 - x_2 - 2x_3$$
$$3x_2 = y_2 - x_1 - x_3 \tag{17}$$
$$2x_3 = y_3 - x_1$$

The procedure is to input a starting set (a_0, b_0, c_0) and get the sequence (a_1, b_0, c_0), (a_1, b_1, c_0), (a_1, b_1, c_1), (a_2, b_1, c_1), etc by using the three equations in turn. Under favourable circumstances the process converges to give the solution (x_1, x_2, x_3) of the equations. Suppose that $a_0 = x_1 \pm \epsilon$, $b_0 = x_2 \pm \epsilon$, $c_0 = x_3 \pm \epsilon$. Starting from error vector $(\pm\epsilon, \pm\epsilon, \pm\epsilon)$, we can see that after using the first equation the error vector is $(\pm\epsilon/4, \pm\epsilon, \pm\epsilon)$. After using the second equation it is $(\pm\epsilon/4, \pm5\epsilon/12, \pm\epsilon)$, which we get by taking the error in b_1 to be $(\pm\epsilon/4)/3 \pm \epsilon/3$, so that the 'worst possible case' error is $\pm5\epsilon/12$. The error vector after the third stage can similarly be calculated to be $(\pm\epsilon/4, \pm5\epsilon/12, \pm\epsilon/8)$. Thus all three errors decrease and repeating the process gives convergence. The key factor in producing convergence is the fact that the dividing number (the diagonal element) is large compared to the elements in each row. For a 2×2 matrix such as

$$\begin{pmatrix} 1 & B \\ C & 1 \end{pmatrix} \tag{18}$$

we can see that after one cycle an error vector $(\pm\epsilon, \pm\epsilon)$ becomes $(\pm B\epsilon, \pm BC\epsilon)$, so that we must get convergence if $|B| < 1$ and $|C| < 1$. In the general $N \times N$ case we *must* get convergence if the sum of the moduli of the off-diagonal elements in each line is less than the modulus of the diagonal element. However, this is not essential; for example, we get convergence in our 2×2 example if $B = \frac{1}{4}$ and $C = 2$. Another case in which the Gauss–Seidel process must converge is when M is symmetric and positive definite (i.e. has all positive eigenvalues).

Because of the simple iterative form of the process I sought for a way to make it work even when M does not satisfy any of the above criteria. I concluded that one systematic way is to proceed from the equation $Mx = y$ to the equation $M^T M x = M^T y$. The matrix $M^T M$ is symmetric and positive definite if M has real elements (which we have assumed in our examples, to avoid complex number arithmetic). Thus the transformed problem must be solvable by the Gauss–Seidel method, with the extra initial expense of two multiplications by M^T. Of course, in this transformed form we could also set

$$x = (M^T M)^{-1} M^T y \tag{19}$$

and proceed by finding the inverse of the square matrix $M^T M$, so that the solution x can be found for any y. $M^T M$ can have an inverse even when M doesn't (e.g. when M is rectangular); in fact for rectangular M the above equation, using an ordinary matrix inverse for $(M^T M)^{-1}$, gives the same result as that obtained using the generalised inverse of M.

3.6 Matrix eigenvalues by iteration

Many of the simple quantum mechanical problems which can be handled on a microcomputer can be treated so that they do not need explicit matrix computations. Even when a matrix diagonalisation is needed it is often possible to use matrices of such a simple form (e.g. tridiagonal) that the eigenvalues can be found without full matrix manipulations (§8.3). However, since I have a soft spot for iterative methods, I will briefly treat an iterative approach which has the extra fascination of using projection operators of a type which have also been used in the quantum theory of angular momentum and in my own work on quantum mechanical applications of finite group theory [7]. The basic idea is very simple. If some matrix (or operator) M has the eigenvalues $(\lambda_0, \lambda_1, \text{etc})$ and a complete set of eigencolumns y_j, then we can express an arbitrary column y in the form

$$y = \Sigma \beta_j y_j. \tag{20}$$

Acting with matrix $M - \lambda \mathbf{1}$ for some λ, with $\mathbf{1}$ the unit matrix, we find

$$(M - \lambda \mathbf{1})y = \Sigma \beta_j (\lambda_j - \lambda) y_j. \tag{21}$$

The multiplying factor $(\lambda_j - \lambda)$ will exactly knock out the y_j component if $\lambda = \lambda_j$; we can make this happen if we know some of the λ_j already. It is clear that after acting many times with $(M - \lambda \mathbf{1})$ we shall have 'shrunk' towards zero the *relative* contribution of all the y_j except that which has the largest $|\lambda_j - \lambda|$ value. Thus the column $(M - \lambda \mathbf{1})^k y$ for sufficiently large k tends towards the y_j with an eigenvalue most remote from λ (in the complex plane if we use a complex M). By inspecting the ratio of the column elements on successive cycles we find $(\lambda_j - \lambda)$. Starting with $\lambda = 0$ initially will give us the eigenvalue $\lambda(\text{max})$ of maximum modulus; setting $\lambda = \lambda(\text{max})$ will then give the eigenvalue at the other extreme of the spectrum. Using the composite multiplier $(M - \lambda_1 \mathbf{1})$ $(M - \lambda_2 \mathbf{1})$, with λ_1 and λ_2 the two known λ values, will lead to some other λ_j. This will probably be the one nearest to the middle of the spectrum, since the multiplying factor is 'cleaning out' the spectrum at equal rates from both ends. However, if the initial column y has 'fluke' large or small values of some β_j, the particular y_j found and the convergence speed can be affected. This calculation, although beautifully simple in concept, requires the operator's skill and experience to make it really effective, and clearly falls into the category of methods which work best when interactive computing is used.

3.7   Matrix folding

I now wish to describe my version of an amusing but useful method which has been used on large computers [8] but will also work on small ones if they can handle arrays. The idea is to convert a matrix eigenvalue problem to a simple single variable problem of the form $E = f(E)$. At least, that is what I do here; most previous authors have simply reduced a large matrix problem to a smaller matrix problem whereas I push the process to the limit of a 1×1 matrix and add a few iterative tricks as well.

The theory can be illustrated by looking at a 3×3 matrix problem with the square matrix and the column being partitioned as shown.

$$\left(\begin{array}{c|c} A & c \\ \hline d & B \end{array}\right)\left(\frac{x}{y}\right) = E\left(\frac{x}{y}\right) \tag{22}$$

A is 2×2, B is 1×1, and so on. Writing out the eigenvalue problem symbolically in the form

$$Ax + cy = Ex$$
$$dx + By = Ey \tag{23}$$

we can solve for y in the second equation and insert the solution in the first equation to give an effective 2×2 problem:

$$[A + c(E - B)^{-1}d]x = Ex \tag{24}$$

x is the projection of the full eigencolumn into the two-dimensional subspace. The last row and column of the 3×3 matrix have been 'folded down' to produce an extra perturbing term in the new 2×2 matrix eigenvalue problem. However, we can further fold down the second row and column of the 2×2 problem to give a 1×1 problem. This will have the x_1 element of x as a common factor on both sides and will be of the form $E = f(E)$. From the 2×2 problem

$$\left(\begin{array}{cc} a & c \\ d & b \end{array}\right)\left(\begin{array}{c} x \\ y \end{array}\right) = E\left(\begin{array}{c} x \\ y \end{array}\right) \tag{25}$$

the reader may verify that the folding down process gives

$$E = a + c(E - b)^{-1}d. \tag{26}$$

This equation is equivalent to a quadratic equation and has two roots, i.e. the two eigenvalues of the original 2×2 matrix. Now, if each element of the 2×2 matrix already involves E, because of folding down from a 3×3 original matrix, then the final equation for E will be equivalent to a cubic one; the three roots will give the three eigenvalues of the original 3×3 matrix.

To pursue the details using algebra is a little messy, but leads to a theory closely related to the Brillouin-Wigner form of perturbation theory [9]. As the reader might have anticipated, on a microcomputer all we do is to invent some simple loop which folds down one row and column *numerically,* and then we repeat the loop until the original matrix is folded down to 1×1 size. The resulting number (1×1 matrix) is some function $f(E)$ of E and when it equals E we have an eigenvalue of the original matrix. The folding rule is quite simple. If the outer row and column to be folded is the Nth, then the resulting $(N-1) \times (N-1)$ matrix has an (m, n) element given by the formula

$$A(m, n) + A(m, N)A(N, n)[E - A(N, N)]^{-1}. \tag{27}$$

This rule can be applied repeatedly until a single number $f(E)$ is obtained. The quantity $E - f(E)$ is then required to be zero. As the folding formula warns us, setting E equal to one of the diagonal elements gives a divergence, and if an eigenvalue is close to one of the $A(N, N)$ then $f(E)$ varies very rapidly with E. One possible program is as follows:

```
10 INPUT Q : DIM A(Q, Q), B(Q, Q)
20 FOR M = 1 TO Q : FOR N = 1 TO Q
25 PRINT M, N : INPUT A(M, N) : NEXT N : NEXT M
30 INPUT E, K
40 FOR M = 1 TO Q : FOR N = 1 TO Q
42 B(M, N) = A(M, N) : NEXT N : NEXT M
50 FOR I = Q TO 2 STEP −1 : D = E − B(I, I)
55 IF D = 0 THEN D = 1E − 8
60 FOR M = 1 TO I − 1 : B = B(M, I)/D
65 FOR N = 1 TO I − 1
70 B(M, N) = B(M, N) + B * B(I, N)
80 NEXT N : NEXT M : NEXT I
90 P = K * E + (1 − K) * B(1, 1)
100 E = P : PRINT P : GOTO 40    (fixed print position)
```

Since the folding process is E-dependent it destroys the original matrix A, so A is kept separately and copied in lines 40 and 42 when required. Lines 50 to 80 do the folding and include a few tricks to cut down the number of matrix element 'look ups' and to reduce the probability of overflow. The resulting 1×1 matrix is $B(1, 1)$, the function $f(E)$ which we require to equal E. The parameter K is a relaxation parameter, so that line 90 forms the quantity

$$P = KE + (1-K)f(E). \tag{28}$$

If $E = f(E)$ then $P = E$ also, but the derivative of P with respect to E can be adjusted to be small even if $f(E)$ has a large derivative:

$$P' = 1 + (1 - K)f'$$
$$= 0 \quad \text{if } K = 1 + f'^{-1}. \tag{29}$$

If f' is -100, for example, then $K = 0.99$ is needed to make P' zero. It is the choice of K to pick out a particular eigenvalue which requires judgment and experience and so is suited to interactive computing. By scanning a wide range of E to find where $E - f(E)$ changes sign the approximate eigenvalue locations can be established. Putting in one of them it is then a matter of adjusting K to achieve convergence. On a computer which allows the operator to use the manual input, for example,

STOP $K = 0.95$ RETURN CONT RETURN

the K value can be changed to control the iterative process. Another 'obvious' approach would be to put the evaluation of $E - f(E)$ as the function evaluation subroutine in a program for Newton's method (§3). However, the function $E - f(E)$ is not smooth, since it has singularities at the $A(n, n)$ values (*except* $A(1,1)$). To produce a smooth function we could multiply it by the product of the $[E - A(n, n)]$ factors. In principle this removes the singularities, although it might not quite do it if rounding errors are present.

A more simple and effective approach is just to exchange the first and the Rth rows and columns of A so that the $A(1, 1)$ element (on to which we do the folding) is the original diagonal element $A(R, R)$ nearest to the eigenvalue which we desire. A simple standard K value of $\frac{1}{2}$ is then usually adequate to give quick convergence, but a preliminary 'shuffling' routine is needed to make B have its rows and columns re-ordered with respect to those of A. The program modification is as follows:

```
30 INPUT E, K, R
44 FOR M = 1 TO Q : B(M, 1) = A(M, R)
46 B(M, R) = A(M, 1) : NEXT M
48 FOR M = 1 TO Q : T = B(1, M)
49 B(1, M) = (R, M) : B(R, M) = T : NEXT M
```

Using the modified program, with $K = \frac{1}{2}$, I looked at the 4×4 matrix with the diagonal elements $A(n, n) = 10n$ and the off-diagonal elements $A(m, n) = 5$. Setting the appropriate R value, so as to 'fold down' on to the nearest $A(n, n)$ each time, I found that it only took a few cycles of iteration each time to give an eigenvalue. The eigenvalues were 45.6149885, 28.9633853, 18.0617754 and 7.35985075. This matrix folding calculation does not give eigencolumns, of course, but putting the good eigenvalues into a matrix projection calculation such as that of §6 would quickly project appropriate eigencolumns out of almost any starting column.

The flowchart for the modified program is as follows:

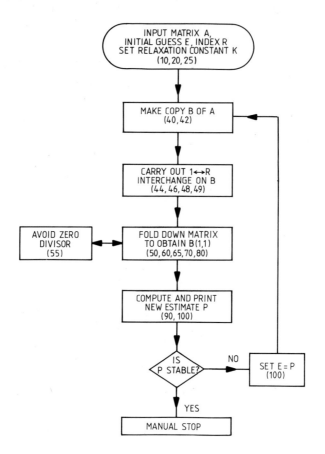

Exercises

1. The result $f(x + h) = f(x) + hf'(x + \theta h)$, with $0 \leqslant \theta \leqslant 1$ for a differentiable function, seems intuitively clear to a physicist, who will think pictorially of the graph of $f(x)$ and see the result. 'Pictorial reasoning' also serves to show that $f(x)$ must be zero somewhere if it tends to $\pm\infty$ at $x = \pm\infty$. The *rule of signs* is an extension of this kind of reasoning. It states that the number of positive roots of a polynomial equation with real coefficients equals S minus an even integer, where S is the number of sign changes along the sequence of coefficients. Devise a pictorial argument to show that if $B^2 < 3AC$ then the cubic equation (7) with real coefficients can have only one real root.

2. Consider what the Newton–Raphson formula looks like if we wish to make $f'(x)$ zero instead of $f(x)$, and see what it becomes when expressed in a simple finite-difference form.

3. Would the lines 100 onwards of our cubic equation BASIC program be of any use in dealing with polynomial equations of other orders?

4. Suppose that the iterative prescription $y = x + \beta f(x)$ gives a first-order process when we are close to a root r, with $y = r + Kh$ if $x = r + h$. If the computer cannot detect the difference between two numbers when they differ by less than 10^{-10}, estimate by how much the 'converged' root obtained can be in error. How does this help to explain the peculiar difficulty with the root of $(x + 1)^3 = 0$ which was noted in the text? Can you see a way of using three successive estimates of the root to produce a good extrapolated value for it?

5. Suppose that we have found a real root r of an Nth degree polynomial equation, $f(x) = 0$. We have $f(x) = (x - r)g(x)$, and $g(x)$ yields an $(N-1)$th degree equation. Show that $g(x)$ can be found by using the process of nested multiplication to work out $f(r)$.

6. Show how to work out $\ln x$ iteratively using a subroutine which can only evaluate e^x.

7. Repeat for a general matrix M the argument used in §3.5 for a 3×3 example and obtain a sufficient condition for convergence of the Gauss–Seidel approach. How could the Gauss–Seidel method be used to find the inverse of M?

8. Write a program which will accept as input the elements of a $Q \times Q$ real matrix and then act repeatedly with $M - \lambda 1$ on a fixed starting column y. If $\lambda - \lambda_j$ is large the elements of the column might become so large that they exceed overflow. Show how to prevent this and at the same time obtain the eigenvalue.

9. On a machine such as the PC-1211 which uses a one dimensional array with elements $A(1)$ onwards, how could the elements of a square matrix be stored and recalled, together with those of several columns? (On the PC-1211 the labels A to Z also serve as names for $A(1)$ to $A(26)$, but we can go beyond 26 stores by using the $A(N)$ notation.)

10. Work out the quantity (dy/dx) for Newton's method and for the simple formula $y = x + \beta f(x)$. Set $x = r$, where r is a root of $f(x) = 0$, and see what you can conclude about the local convergence properties of the two methods.

11. To a casual reader the set of eigenvalues quoted in §3.7 for the 4×4 matrix might look reasonable but have to be taken on trust. Can you think of any simple ways of testing them without actually doing the full calculation?

12. Can you work out what the function $f(E)$ of §3.7 looks like if it is written out as a series of terms involving E and the matrix elements? Try a 2×2 and a 3×3 example to see the form of the series.

Solutions

1. If the $f(x)$ curve crosses the axis three times there will be two points at which $f'(x) = 0$. The condition that the equation $3Ax^2 + 2Bx + C = 0$ shall have two real roots is $B^2 > 3AC$, so this is a necessary (but not sufficient) condition that the cubic equation has three real roots.

2. We have

$$y = x - f'(x)/f''(x) \tag{30}$$

Using subscripts 0, ± 1 for x, $x \pm h$ and taking the lowest order finie-difference forms for the derivatives gives

$$y = x - \frac{h}{2}(f_1 - f_{-1})/(f_1 - 2f_0 + f_{-1}). \tag{31}$$

This is the same result as found using the Newton–Gregory formula in solution 5.2. If we do not know (or cannot bother to calculate) any other f values then we have to stop after the first iterative cycle. To continue we have to work out three new f values.

3. If $N - 2$ roots of an Nth degree equation have been found the procedure can be modified slightly to give the last two roots. All we have to do is to remember that for the equation $A_N x^N + \ldots + A_0 = 0$ the N roots have the sum $-A_{N-1}/A_N$ and the product $(-1)^N A_0/A_N$. If the first $N - 2$ roots are real they could be located by varying the input x_0 in the Newton–Raphson method.

4. The computer declares convergence when $y = x \pm \epsilon$, where ϵ is the tolerance level (either intrinsic or declared in the program). If the input x varies by amount Δx then $(y - x)$ varies by amount $(K - 1)\Delta x$. Thus if we are within distance $\delta = \epsilon/(K - 1)$ of the root r the computer will see $y - x$ as effectively zero and convergence will apparently be obtained. By varying x_0 we will vary the estimate of r throughout a band of width roughly 2δ. This analysis implies that a slowly converging one-sided process (with $K \cong 1$) gives most problems, and our case $(x + 1)^3 = 0$ involves a K value which, although not constant, is very close to 1 as the root -1 is approached. The problem arises from the use of the finite-difference form for $f'(x)$. Although this has the great advantage of making the program 'universal' it causes trouble here. If we explicitly put the analytical form of $f'(x)$ i.e. $3(x + 1)^2$ into the program, we get, at $x = -1 + h$,

$$y = -1 + h - h^3/(3h^2) = -1 + \tfrac{2}{3}h. \tag{32}$$

We thus have a converging first-order process yielding an accurate root, although it involves a longer running time than the normal second-order process for separate roots.

For any equation, an exact first-order process is such that if we know $x_0 = r + h$, $x_1 = r + Kh$, $x_2 = r + K^2h$ then the errors form a geometric progression and the result

$$r = [x_0 x_2 - x_1^2]/[x_0 - 2x_1 + x_2] \tag{33}$$

holds. The numerator here equals r times the denominator, with a common factor $(1-K)^2$. An alternative form which is not so subject to loss of accuracy when K is close to 1 or h is very small is

$$r = x_0 - (x_1 - x_0)^2/[x_0 - 2x_1 + x_2]. \tag{34}$$

These formulae are often used to accelerate the converge of a sequence $\{x_n\}$ by forming $r(0, 1, 2), r(1, 2, 3)$ and so on and looking for convergence of the r sequence. The method is then usually called the *Aitken δ^2 process* and is actually a special case of the use of Padé approximants (§6.3); if each x_N is the sum up to the x^N term of a power series, then the rs are $[N-1/1]$ approximants. For obvious reasons this approach is often called the *geometric approximation* by physicists. Its most fascinating applications occur for divergent sequences with $K > 1$; the formula for r is still valid for our postulated sequence and r is sometimes called the *anti-limit* of the divergent sequence. More complicated divergent sequences can also yield convergent $\{r\}$ sequences. An interesting article by Shanks [10] gives many applications of the Aitken and Padé procedures.

5. The proof of the method simply involves writing down the equation $f(x) = (x - r)g(x)$ and equating coefficients of powers of x. The resulting recurrence relation is then the same as that for the successive terms in the nested multiplication procedure. We give the example $f(x) = 4x^3 + 5x^2 - 6x - 7$, for which we have the root -1 from the earlier examples. Working out $f(-1)$ we find, with $x = -1$,

$$(4x) + 5 = 1$$
$$(1x) - 6 = -7 \tag{35}$$
$$(-7x) - 7 = 0$$

so that the other roots obey the equation $4x^2 + x - 7 = 0$. The coefficients are obtained from the round brackets in the nested multiplication. Clearly we could get the real root r of a cubic equation and then get a quadratic equation for the other roots; the program in the text proceeds without explicitly finding the coefficients.

6. To get $\ln N$ we must solve the equation $e^x - N = 0$.
Newton's method gives the input-output prescription

$$y = (x - 1) + Ne^{-x}$$

which converges to ln N if we set $x_0 = 1$. We don't need to use N values greater than e.

7. The first equation of the Gauss–Seidel procedure will be

$$M_{11}x_1 = y_1 - \Sigma M_{1j}x_j \tag{36}$$

where the summation excludes M_{11}. Suppose that the exact solution has components X_j, denoted by a capital letter. Then if the initial guess has the elements $x_j = X_j + \epsilon_j$ we obtain

$$M_{11}x_1 = (y_1 - \Sigma M_{1j}X_j) - \Sigma M_{1j}\epsilon_j$$
$$= M_{11}X_1 - \Sigma M_{1j}\epsilon_j. \tag{37}$$

If the largest of the ϵ_j has modulus $|\epsilon|$ then the x_1 value obtained, $X_1 + \epsilon_1$, must be such that the following inequality is obeyed:

$$|\epsilon_1| \leqslant |\epsilon|\,|M_{11}|^{-1}\,\Sigma\,|M_{1j}| \tag{38}$$

Proceeding down the set of equations we can see that convergence must result if the multiplying factor of $|\epsilon|$ is less than one on each line, so we converge to the solution by going through many circuits of the equations. The use of the *modulus* of the various quantities shows that the results still hold for complex matrices. If M is Hermitian, the diagonal elements are real and so only complex multiplication (not division) is needed to apply the Gauss–Seidel procedure. To obtain the jth column of the inverse matrix M^{-1} we simply solve $Mx = y$ for a column y with 1 in the jth position and zeros in all other positions.

8. A possible program (in Pet style) is as follows

```
  5 Q = 3
 10 DIM M(Q, Q), X(Q), Y(Q), Z(Q)
 20 FOR M = 1 TO Q : PRINT M : INPUT X(M)
 30 FOR N = 1 TO Q : PRINT M, N : INPUT A(M, N)
 40 NEXT N : NEXT M : PRINT "S" : INPUT S
 50 FOR M = 1 TO Q : Y = 0
 60 FOR N = 1 TO Q
 70 Y = Y + A(M, N) * X(N) : NEXT N
 80 Z(M) = Y - S * X(M)
 90 NEXT M
100 Z = Z(1) : FOR M = 1 TO Q - 1
110 IF ABS (Z(M+1)) > ABS (Z) THEN Z = Z(M+1)
120 NEXT M : FOR M = 1 TO Q
130 X(M) = Z(M)/Z : NEXT M : PRINT Z : GOTO 50
```

The dimension statement on line 10 is needed on many microcomputers.

Lines 20 to 40 accept the arrays M and x and the parameter S for the matrix $M - S1$. Lines 50 to 80 work out $(M - S1)x$. Lines 100 to 130 display the element of largest modulus in the x column (the number which eventually converges to the eigenvalue), then set it equal to 1 and scale the other elements proportionately (thus avoiding overflow). Lines 50 to 80 include a 'clever' feature which is useful on some microcomputers. To look up a subscripted variable such as Y(M) takes longer than looking up a simple variable Y in the storage locations. By using Y instead of Y(M) in the matrix multiplication and setting Y(M) = Y *once* at the end I found that a time saving of 25 per cent in matrix multiplications is obtained on a Pet (of 2001 series type). Perhaps some reader will come up with an even shorter procedure!

9. If we are studying a $Q \times Q$ matrix then we can use the description $A(M * Q + N)$ for the element $A(M, N)$ of a square matrix. The PC-1211 will accept and operate with such a labelling scheme. If we wanted to have two Q-colums X and Y as well we could use the translation $X(N) \rightarrow A(Q * Q + N)$, $Y(N) \rightarrow A(Q * Q + Q + N)$, and so on, the Ith column being denoted by $A(Q * Q + Q * I + N)$. The numbers $Q * Q$ and $Q * Q + Q$ can be written as constants, of course, when Q has been fixed. On a TI-58 it is necessary for the operator to designate the specific location of each element, which involves a full writing out of at least part of each matrix multiplication, although clever tricks with the indirect addressing facility can probably give some shortcuts.

10. The (dy/dx) value for Newton's method is $[f(x)f''(x)]/[f'(x)]^2$; if this quantity has a modulus greater than 1 the process will diverge, but if x_0 is sufficiently close to r the $f(x)$ factor will make (dy/dx) small. For the simple formula the (dy/dx) value is $1 + \beta f'(r)$, so that the condition for convergence is

$$-1 < 1 + \beta f'(r) < 1 \tag{39}$$

which means that $\beta f'(r)$ must be between 0 and -2 to get local convergence to r. In principle then by varying β and x_0 we can 'tune in' to various roots in an interactive microcomputer calculation. One procedure which is some-times used in iterative procedures is the *relaxation* procedure (which is related to the relaxation methods used for solving differential equations and linear equations). If the iterative input-output formula has input x and output $y(x)$ then the relaxation procedure takes a weighted average of the input and output. The resulting estimate takes the form

$$z(x) = \alpha x + \beta y(x)$$

$$= x + \beta[y(x) - x] \tag{40}$$

since we must have $\alpha + \beta = 1$ to ensure that $z(r) = y(r) = r$. $\beta > 1$ gives *over-relaxation*, with the shift distance $y(x) - x$ being increased over its normal ($\beta = 1$) value; $\beta < 1$ gives *under-relaxation*. Since $z'(r) = (1 - \beta) + \beta y'(r)$ it is in principle possible to choose β to increase the rate of convergence or even with care to render a divergent procedure convergent. Simple over-relaxation procedures have sometimes been used to speed up convergence of the Gauss–Seidel procedure (§5) which solves a set of simultaneous linear equations. My own analysis of that problem suggests that only by using a complicated *matrix* version of the parameter β could we convert a divergent Gauss–Seidel calculation into a convergent one, so the easiest way to make the Gauss–Seidel method 'universal' is to use the transpose matrix as I suggest in the text. There is, of course, another way in which we could try to use a divergent sequence of estimates to a root; we could 'Aitkenise' it (solution 4). This procedure often works if the initial estimate x_0 is close to r. Since the program in the text for a cubic equation hasn't failed so far I suspect (although I have been too lazy to check) that Newton's method must always converge for at least one real root of a cubic equation!

11. The most simple procedure is to use the generalised diagonal sum rule, which says that the sum of the nth powers of the eigenvalues equals the sum of the diagonal elements of the nth power of the matrix. The diagonal sum is 100 for the matrix treated in the text. To find the diagonal sum of A^2 only requires calculation of the four diagonal elements. The sum is quickly found to be 3300. The sum of the quoted eigenvalues is $100 - (5E - 8)$ and the sum of their squares is $3300 - (2E - 6)$, as evaluated on the PC-1211. Pretty good!

12. The result for $f(E)$ is

$$A(1, 1) + \Sigma A(1, n)A(n, 1) d(n)$$
$$+ \Sigma \Sigma A(1, n)A(n, m)A(m, 1) d(n) d(m) + \ldots \qquad (41)$$

Here $d(n)$ denotes $[E - A(n, n)]^{-1}$ and the summations exclude the index 1 for n or m. The chain-like structure of the terms is exactly maintained in the later terms. For a 3×3 matrix the later terms do not contribute and it is easy to see how the equation $E = f(E)$ is equivalent to a cubic equation. In perturbation theory the series above is known as the Brillouin–Wigner eigenvalue series, when written in the form $E = f(E)$. The terms in $f(E)$ can be cleared of E if E in the denominators is replaced by the $f(E)$ expansion. By turning the series inside out in this way we obtain the more complicated Rayleigh–Schrödinger series for E, although in my use of Rayleigh–Schrödinger theory in this book (§§9.2 and 12.4) I have avoided the sum-over-states language for expressing the coefficients in the E series. If

$A(1, 1)$, the analogue of the first-order energy, is zero, then the BW and RS series are the same up to the third order. For a symmetric matrix, stopping at the third-order term in the above series necessarily gives an upper bound to the lowest eigenvalue if the equation $E = f(E)$ is solved. This is one of the useful theorems from BW theory.

When an $n \times n$ submatrix of an $N \times N$ matrix is being diagonalised, one widely used approximation is to allow in lowest order for the effect of the other $(N - n)$ rows and columns by adding a sum of terms of type $A(m, k)A(k, n)d(k)$ to the (m, n) element of the $n \times n$ submatrix. If the effect is small then it can be taken to give a reasonable simulation of the effect of a full matrix folding from $N \times N$ down to $n \times n$, and the E in $d(k)$ can be replaced by $A(1, 1)$, the 'diagonal energy', to avoid having E implicitly appearing in the matrix elements. I discuss this kind of procedure in my review article and book [7, 9].

Notes

1. I M Longman 1960 *Math. Comput.* **14** 187
2. G Joos 1958 *Theoretical Physics* (London & Glasgow: Blackie) p 100
3. D R Green and J Lewis 1978 *Science with Pocket Calculators* (London: Wykeham Publications) p 186
4. J C Slater 1939 *Introduction to Chemical Physics* (New York: McGraw-Hill) Ch 12
5. A Ben-Israel 1966 *Math. Comput.* **20** 439
6. A Mackay Sept. 1981 *Practical Computing* p 108
7. J P Killingbeck 1975 *Techniques of Applied Quantum Mechanics* (London: Butterworths)
8. A R Williams and W Weaire 1976 *J. Phys. C: Solid State Phys.* **9** L47
9. J Killingbeck 1977 *Rep. Prog. Phys.* **40** 963
10. D Shanks 1955 *J. Math. Phys.* **34** 1

4 Some finite-difference methods

4.1 Introduction

Almost any book on numerical analysis will contain a great deal of material on finite-difference methods. In this short chapter I deal with a few simple formulae which are needed throughout the book in the setting up of microcomputer programs and in the analysis of the output of the programs. Already in §3.3 I have used a simple finite-difference formula for a derivative; in §4.3 I give some more accurate formulae for first and second derivatives which are used in chapter 10 to treat the Schrödinger equation. The Richardson extrapolation method explained in §4.3 is used repeatedly throughout the book, particularly in chapters 5, 10 and 12. The simple Newton–Gregory formula of §4.2, although not an 'advanced' formula in the hierarchy of numerical methods, is quite adequate for handling interpolations with the smooth functions which appear in the quantum mechanical calculations of this book. I throw in the Euler transformation (§4.2) as an example of one way of taming a poorly convergent series; this problem is looked at in more detail in chapter 6.

To state the case briefly: a computer does things with numbers, but we want to get it to handle problems concerning differentiation and integration, about which it 'knows' nothing. Finite-difference methods represent one way to translate problems of the calculus into discrete numerical problems which the computer can handle. Chapters 5, 10 and 12 provide plentiful examples of this translation process in action.

4.2 The Newton–Gregory formula

Finite-difference methods are often used on computers to convert problems involving differential equations into problems involving recurrence relations or matrix calculations. I use only a few basic formulae in this book, but it is interesting to note how the operator algebra employed in quantum mechanics is also useful in finite-difference work. One approach is to start from the conventional Taylor series expansion and note that it is formally an exponential involving the differentiation operator D;

$$f(x + h) = f(x) + h \, Df(x) + \tfrac{1}{2}h^2 \, D^2 f(x) + \ldots$$

$$= e^{hD}f(x) = E(h)f(x) \tag{1}$$

where $E(h)$ is the *shift operator* which converts $f(x)$ to $f(x + h)$. The *forward-difference* operator $\Delta(h)$ is defined by

$$\Delta(h)f(x) = f(x + h) - f(x)$$

$$= E(h)f(x) - f(x). \tag{2}$$

If $f(x)$ is tabulated at intervals of h, then the $\Delta(h)f$ values are in the first column of differences. Conventional treatments use the standard interval $h = 1$. I shall keep h explicitly for clarity in some formulae, but shall often use obvious abbreviations such as f_0, f_1, f_{-1} for the quantities $f(x), f(x + h), f(x - h)$.

The equations above lead to the formal result

$$E(h) = 1 + \Delta(h) = e^{hD} \tag{3}$$

from which various equations follow by manipulation. For example

$$hD = \ln (1 + \Delta)$$

$$= \Delta - \tfrac{1}{2}\Delta^2 + \tfrac{1}{3}\Delta^3 \ldots \tag{4}$$

The quantities $\Delta^2 f$, $\Delta^3 f$, etc, appear in the second and third difference columns when f is tabulated at interval h, as shown below

	f	Δf	$\Delta^2 f$
$x - h$	f_{-1}		
x	f_0	Δf_{-1}	$\Delta^2 f_{-1}$
$x + h$	f_1	Δf_0	$\Delta^2 f_0$
$x + 2h$	f_2	Δf_1	

Here $\Delta f_0 = f_1 - f_0$ and Δ^2 is interpreted in the usual linear operator spirit:

$$\Delta^2 f_0 = \Delta(\Delta f_0)$$

$$= \Delta(f_1 - f_0) = \Delta f_1 - \Delta f_0$$

$$= f_2 - 2f_1 + f_0. \tag{5}$$

Another result obtainable from the formulae above is as follows:

$$f(x + Nh) = e^{NhD}f(x)$$
$$= (1 + \Delta)^N f(x)$$
$$= [1 + N\Delta + \tfrac{1}{2}N(N-1)\Delta^2 + \ldots]f(x). \tag{6}$$

This result is often called the Newton–Gregory forward-difference interpolation formula, and allows us to obtain interpolated values for f at x values between those shown in the table. Taking terms up to $\Delta^2 f_0$, for example, clearly is using the values of f_0, f_1, f_2 and is equivalent to fitting a second-order polynomial in x to these three (f, x) pairs. (The usual procedure of Lagrange interpolation would yield the same result.)

The *Euler transformation,* which is useful in the summation of real alternating series, can also be written to involve forward-differences in a table which displays the coefficients of the series. We start from the series

$$f(z) = \Sigma (-1)^n A_n z^n \tag{7}$$

and multiply by $(1 + z)$ to get

$$(1 + z)f(z) = A_0 + (A_0 - A_1)z + \ldots$$
$$= A_0 - z \Sigma (-1)^n \Delta A_n z^n. \tag{8}$$

Now we can repeat the procedure on the new alternating series with coefficients ΔA_n, and so on in an iterative analytic process which will yield the formal result

$$f(z) = (1 + z)^{-1} \Sigma (-y)^n \Delta^n A_0 \tag{9}$$

with $y = z(1 + z)^{-1}$. It is clear that as z varies from 0 to ∞ through real values the transformed variable y varies from 0 to 1. Similar transformations for reducing an infinite range to a finite one are sometimes used in numerical integration ($\S5.7$) and in finite-difference simulations of differential equations (chapter 10). The inverse transformation is $z = y(1 - y)^{-1}$.

4.3   Derivatives and Richardson extrapolation

Besides the forward-difference operator Δ we can also use the *backward-difference* operator ∇, defined by the formula

$$\nabla(h)f(x) = f(x) - f(x - h) \tag{10}$$

and obeying the relationship

$$E^{-1}(h) = 1 - \nabla(h) = e^{-hD}. \tag{11}$$

Another operator, actually the most useful one for getting accurate finite-difference formulae, is the *central-difference* operator, $\delta = e^{1/2hD} - e^{-1/2hD}$ Squaring δ and remembering the relationship between the various operators we get

$$\delta^2 f_0 = f_1 - 2f_0 + f_{-1}$$
$$= h^2 D^2 f_0 + \tfrac{1}{12} h^4 D^4 f_0 + \dots \tag{12}$$

a result which is much used in finite-difference treatments of the Schrödinger equation (§10.2). The result follows from the Taylor series, of course, without the use of the δ formalism. Indeed, we can take a head-on approach and simply construct linear combinations of the values $f(x)$, $f(x \pm h)$, etc, which agree with $Df(x)$ and $D^2 f(x)$ to some specified order of h. One smart way to do this is as follows. The first obvious approximation to Df_0 is

$$Df_0 = (f_1 - f_{-1})/2h \tag{13}$$

and the Taylor series shows that it is in error by a leading term of order h^2. If we use $2h$ instead of h, then, we multiply the error by 4, so we *remove* the h^2 error by mixing the results for h and $2h$ in a ratio of $4:-1$. This idea quickly gives us

$$Df_0 = (f_{-2} - 8f_{-1} + 8f_1 - f_2)/12h \tag{14}$$

which has an error of order h^4. In computer work it is not really essential to derive the second formula above; we can just work out the two estimates $(f_1 - f_{-1})/2h$ and $(f_2 - f_{-2})/4h$ as numbers and take the 4 to -1 mixture *numerically*. In analytical work we like to have a formula for everything, but it is often easier to just let the computer do the relevant manipulations numerically, arriving at the result which it would have got if we had solved all the equations analytically and put the solution formulae into the program. Proceeding as above we can also obtain two formulae of increasing accuracy for the second derivative:

$$D^2 f_0 = (f_{-1} - 2f_0 + f_1)/h^2 \tag{15}$$
$$D^2 f_0 = (-f_{-2} + 16f_{-1} - 30f_0 + 16f_1 - f_2)/12h^2. \tag{16}$$

A shorthand way of referring to a linear combination such as this last one is by the symbol $(-1, 16, -30, 16, -1)$ with the f_0 coefficient placed centrally. The coefficients have a zero sum, since a constant function has zero derivatives. The denominator can be deduced by noting that the function x^2 has $D^2 f_0 = 2$.

In various processes of numerical integration (§§5.3, 10.2) we obtain values of a quantity (e.g. an integral or an energy value) at intervals h. We thus start with $f(h)$, $f(2h)$, etc, and often want to know $f(0)$. Extrapolating by fitting the f values to a polynomial in h is wasteful, since we already know from theory (for the class of methods used in those sections) that f is a function of x^2. We can thus assume that

$$f(x) = A_0 + A_2 x^2 + A_4 x^4 + \ldots \tag{17}$$

Setting $x = h$, $2h$, etc into this postulated equation gives equations from which the A_n can be eliminated in turn. For example,

$$f(h) = A_0 + A_2 h^2 + A_4 h^4 + \ldots$$
$$f(2h) = A_0 + 4A_2 h^2 + 16A_4 h^4 + \ldots \tag{18}$$

from which we conclude that

$$\tfrac{1}{3}[4f(h) - f(2h)] = A_0 - 4A_4 h^4 + \ldots \tag{19}$$

which gives an improved estimate of A_0. Using three f values we can eliminate A_4, and so on. This approach is often called *Richardson extrapolation* in the literature; when the fs are the values of integrals obtained using strip widths h, $2h$, etc it is often termed *Romberg integration*. Romberg used the particular sequence h, $\tfrac{1}{2}h$, $\tfrac{1}{4}h$, etc, in his work on integration. Working out the linear combination of the $f(nh)$ which eliminates as many A_n as possible is a once-for-all piece of arithmetic and I collect the results in a table below. If, for example, we have f_1, f_2 and f_4 (in an obvious shorthand notation) then we get the best f_0 estimate by taking the combination $64f_1 - 20f_2 + f_4$ and dividing by 45.

1	2	3	4	D
4	−1	−	−	3
64	−20	−	1	45
15	−6	1	−	10
56	−28	8	−1	35

Exercises

1. Consider the triangle problem below, pointed out to me by Mr Barry Clarke when he was an undergraduate student at Hull. Suppose that n layers of small equilateral triangles have been set out as shown. At $N = 3$, say, there

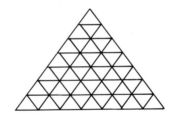

are actually $U(3) = 10$ upwards-pointing triangles, since there are triangles of height 1, 2 and 3 units. We have here an escalating combinatorial problem if we want to know $U(N)$ for general N. Make a table of the $U(N)$ values by explicit counting and use the Newton–Gregory formula of §2 to get a formula for $U(N)$.

2. Suppose that a function $f(x)$ has a minimum at $x = m$. We have located m roughly and have the three numbers $f(a)$, $f(a \pm h)$, where $a + h > m > a - h$. How can we get a closer estimate of m using the three f values?

Solutions

1. The table and its differences are as follows.

N	U	ΔU	$\Delta^2 U$
0	0		
1	1	1	
2	4	3	2
3	10	6	3
4	20	10	4
5	35	15	5
6	56	21	6

In this case the interval h is 1. Without completing the table fully we can see that $\Delta^3 U = 1$ and $\Delta^4 U = 0$ throughout. This shows at once that $U(N)$ is a polynomial of degree 3. From the Newton–Gregory formula at $N = 0$ we have

$$U = 0 + N1 + \tfrac{1}{2}N(N-1)2 + \tfrac{1}{6}N(N-1)(N-2)1$$

$$= (N^3 + 3N^2 + 2N)/6. \tag{20}$$

The functions $N(1) = N$, $N(2) = N(N-1)$, $N(3) = N(N-1)(N-2)$, etc, have the property that $\Delta N(M) = MN(M-1)$ which is analogous to the property $D(x^M) = Mx^{M-1}$ in the differential calculus. This highlights the close similarity between the Newton–Gregory expansion using the $N(M)$ and the Taylor expansion using a power series.

2. From the three values, f_{-1}, f_0 and f_1, say, we can form two differences (Δf_{-1} and Δf_0) and one second difference ($\Delta^2 f_{-1}$). We then have

$$f(-h + Nh) = f_{-1} + N\Delta f_{-1} + \tfrac{1}{2}N(N-1)\Delta^2 f_{-1}. \tag{21}$$

N is here a continuous variable to permit interpolation, so we can set df/dN equal to zero to estimate the location of the minimum. We find

$$\mathrm{d}f/\mathrm{d}N = \Delta f_{-1} + (N - \tfrac{1}{2})\Delta^2 f_{-1} = 0. \tag{22}$$

The minimum is at $a - h + Nh = a + (N-1)h$. A little algebra gives

$$m = a + \frac{h(f_{-1} - f_1)}{2(f_{-1} + f_1 - 2f_0)} \tag{23}$$

as the explicit formula for m, although, of course, we could proceed by simply using the numerical values of the differences. What we are doing is to construct a parabolic interpolating curve for f and then compute the minimum using the curve. If we wanted to get an even better result we could next evaluate $f(m)$, $f(m \pm h/2)$ and do the analysis again. The analysis outlined here is needed in the calculation of resonant state energies using finite-difference methods ($\S 12.5$).

5 Numerical integration

5.1 Introduction

In this chapter I deal with a few topics from the subject of numerical integration. An expert on numerical integration techniques would undoubtedly object to my neglect of the various forms of Gaussian integration, and I have not attempted to give a full view of all possible methods. I have simply chosen some useful methods (mainly of what is usually called the Newton–Cotes type) which also lend themselves to easy analysis by means of the theoretical tools which I employ in this book. The analysis in §§5.3 and 5.4 uses simple ideas about Taylor series and Richardson extrapolation and also involves asymptotic series, which get a more detailed treatment in chapter 6. In carrying out numerical integration on a microcomputer it is important to take care over any special defects which a particular machine may have in its arithmetic operations, so reference back to the calibratory tests of chapter 2 is sometimes made in this chapter. In §5.7 two problems involving endpoint singularities are shown to arise from anharmonic oscillator problems in classical and quantum mechanics. At several points I illustrate the relevance of my comment in chapter 1, that the best way to write a formula for analytical work is not always the best way to write it when it is to be used in a microcomputer program.

5.2 A Simple test integral

The most simple integration rule is the *midpoint rule,* which approximates an integral by using N strips of width h to cover the integration region and gives

$$I = \int_A^B y(x)\,dx \cong \sum_{j=0}^{N-1} hy(x_j) \tag{1}$$

with $x_j = (j + \frac{1}{2})h$ and $h = (B - A)/N$. (For an infinite convergent integral we give h directly, since N is not defined unless we impose a cut-off at some finite distance.) The simple midpoint rule has the advantage that it gives a well-defined result even if the integrand is one like $(x - A)^{-1/2}$ which has a singularity at $x = A$. To obtain what in calculus is called *the* value of the integral we should really take the limit $h \to 0$, and one approach sometimes used is to compute the integral several times, repeatedly halving the stripwidth until the estimate obtained is stable to the number of digits required. This approach is usually very slow compared with the Romberg approach which I describe below, and there are two obvious reasons why the limit $h \to 0$ is not directly attainable. First, taking $h \to 0$ directly gives a running time tending to infinity. Second, adding up so many numbers gives a rounding error which will make the apparent limit differ from the true one. To attain the limit $h \to 0$ we have to use an indirect approach blending analysis with computation. To start off the discussion I give below a table showing values of the integral of x^4 between 0 and 1 as computed using the midpoint rule on a TI-58. (From my earlier results (§§2.2 and 2.3) it follows that for the Pet it is wise to use $X * X * X * X$ instead of $X \uparrow 4$ and also to move from x to $x + h$ by using $N = N + 1 : X = N * H$. Further, to reduce rounding error, the sum of the f values, S, is formed, I being taken as the product hS.)

h	0.100	0.050	0.025
I	0.198 336 25	0.199 583 5157	0.199 895 8448
$\epsilon(h)h^{-2}$	0.166 375	0.166 593 72	0.166 648 32

The exact value is $\frac{1}{5}$, so we can work out the error ϵ exactly for this simple test case. The results suggest that ϵ is closely proportional to h^2 for small h and so stimulate us to investigate this analytically. Taking the result as an empirical discovery for the time being we can extrapolate to see what the answer at $h = 0$ would be. The results at $h = 0.05$ and $h = 0.025$ give the extrapolated estimate 0.199 999 9545, with an error of 5×10^{-8}. To achieve this error with a direct calculation would need an h value of about $(5 \times 10^{-8}/0.168)^{1/2} = 0.0005$, taking fifty times as long as the calculation at $h = 0.025$, and, of course, needing calculations at intermediate h values to check that convergence is being approached.

5.3   A Taylor series approach

I now work out formally the area in a single strip, taking the origin $x = 0$ to be at the centre and expanding the integrand as a Taylor series:

$$\int_{-h/2}^{h/2} y(x) \, dx = \int \left[y_0 + \frac{1}{2} x^2 D^2 y_0 + \frac{1}{24} x^4 D^4 y_0 \ldots \right] dx. \tag{2}$$

The odd powers of x are omitted because they contribute nothing to the integral. The result is

$$\int y(x) \, dx = h y_0 + \frac{1}{24} h^3 D^2 y_0 + \frac{1}{1920} h^5 D^4 y_0 \ldots \tag{3}$$

Adding together the area of all these trips between the integration limits A and B we get a midpoint sum plus correction terms:

$$\int_A^B y(x) \, dx = \sum_j \left[h y_j + \frac{1}{24} h^3 D^2 y_j + \frac{1}{1920} h^5 D^4 y_j \ldots \right]. \tag{4}$$

However, applying the same approach to the integral of $D^2 y$ and $D^4 y$ we find

$$\int_A^B D^2 y(x) \, dx = \sum_j \left[h D^2 y_j + \frac{1}{24} h^3 D^4 y_j + \ldots \right] \tag{5}$$

$$= Dy(B) - Dy(A). \tag{6}$$

$$\int_A^B D^4 y(x) \, dx = \sum_j h D^4 y_j + \ldots \tag{7}$$

$$= D^3 y(B) - D^3 y(A). \tag{8}$$

Putting all the pieces together gives

$$\int_A^B y(x) \, dx = h \sum y_j + \frac{1}{24} h^2 \, [Dy]_A^B - \frac{7}{5760} h^4 \, [D^3 y]_A^B \ldots \tag{9}$$

Note how I used a simple 'iterative' argument here, in keeping with my general approach throughout the book. The resulting formula (9) is often called an Euler–Maclaurin formula in the literature. The *trapezoidal rule* starts from the prescription

$$\int_A^B y(x) \, dx = \frac{1}{2} [y_0 + y_N] h + \sum_{j=1}^{N-1} h y_j \tag{10}$$

with $y_j = jh$, and includes the endpoints B and A in the sum. The associated Euler–Maclaurin formula in this case is

$$\int_A^B y(x) \, dx = I(\text{trapezoidal}) - \frac{1}{12} h^2 \, [Dy]_A^B + \frac{3}{2160} h^4 \, [D^3 y]_A^B \ldots \tag{11}$$

A glance at these results shows that the midpoint rule is more accurate than the trapezoidal one, for sufficiently small h. On looking through the textbooks of numerical analysis I found that the midpoint rule is treated much less often than the trapezoidal rule. Several authors briefly comment that closed Newton–Cotes formulae (i.e. those using endpoint values) are in general more accurate than open Newton–Cotes formulae, without pointing out that this is not so for these two founder members of each family. By combining the midpoint and trapezoidal estimates in a ratio of 2 to 1 we remove the h^2 error term and leave a leading error term proportional to $h^4[D^3y]$. The particular sum of terms involved is then

$$\frac{2}{3}[y_{1/2} + y_{3/2} + \ldots]h + \frac{1}{3}\left[\frac{1}{2}y_0 + y_1 + y_2 + \ldots\right]h \tag{12}$$

$$= \frac{1}{2}h[y_0 + 4y_{1/2} + 2y_1 + 4y_{3/2} + \ldots]/3 \tag{13}$$

which is Simpson's rule for stripwidth $\frac{1}{2}h$: end values plus 2× (even values) + 4× (odd values), times $h/3$. The Simpson's rule error for stripwidth h has the leading term $-\frac{1}{180}h^4[D^3y]_A^B$. It is clear from the treatment above, and by dimensional analysis, that the terms in the series are of form $h^{n+1}[D^ny]$ with n odd. This suggests (correctly) that Simpson's rule is exact for any polynomial of degree three or less (see §1.3). It also seems to imply that the simple rules give exact values for the integral if the integrand $y(x)$ is periodic with period $(B-A)$, since then all the terms of form $[D^ny]$ vanish. For $A=0$, $B=\infty$, it seems that the integral of functions such as $\exp(-x^2)$ or $(1+x^2)^{-1}$ should be given exactly by the simple integration rules, since they have all their D^ny values (for odd n) equal to zero at 0 and ∞. Leaving the periodic function case to exercise 3, I have computed the value of the integral of $\exp(-x^2)$ between 0 and ∞ on the PC-1211, using varying stripwidth h and the midpoint rule. The exact value is $\frac{1}{2}\sqrt{\pi}$, or 0.886 226 925 45 to eleven digits. The midpoint rule results give a value of 0.886 226 925 45 up to the remarkably large stripwidth $h = 0.6$, but for larger h the result falls away rapidly from the true result. The sums for $(1+x^2)^{-1}$ take a long time to evaluate but in any case we can see that for large h the trapezoidal rule sum must vary as h^{-1} and so cannot equal the correct value of the integral for all h. The point is that the Euler–Maclaurin series is an asymptotic one, such as the ones discussed in chapter 6, so that only in the limit $h \to 0$ does it give an exact estimate of the error. The detailed theory for the special cases in which the series vanishes has been discussed by several authors [1, 2].

5.4 Romberg integration

In §4.3 I gave a table of the extrapolating formulae used for calculating $I(0)$ when $I(n_1 h)$, $I(n_2 h)$ etc are known for integer n_1 and n_2 and when the difference $I(h) - I(0)$ is known to be a series in h^2. This covers the case of the Euler–Maclaurin formulae, if we neglect the residual error $R(h)$ missed by the series. The examples above show that $R(h)$ is non-zero; however, it tends to zero with h more quickly than any positive power of h. The name *Romberg integration* nowadays is used to describe any integration process which uses a set of estimates $I(n_j h)$ to find an extrapolated I value. Romberg's original process uses $h, \frac{1}{2}h, \frac{1}{4}h$, etc (a halving process). It will converge to the correct value (given negligible rounding error) even for cases with a non-zero residual error term, because for $h \to 0$ that term is always swamped by the power series terms. Although the Euler-Maclaurin formulae mention only derivatives at A and B, the derivation assumed that the functions concerned are smooth everywhere between A and B, so that the final result is deceptively simple. If the integrand y has a kink at $x = (A + B)/2$, say, then we can arrange that the kink falls at a strip boundary. Applying the theory twice, once to each half of the range, we get two extra terms involving the left and right derivatives at the kink. Nevertheless, the series still involves h^2 and can be treated numerically by the same kind of extrapolation calculation as used for a smooth function. For the integral of x^4 treated earlier an h^2 law extrapolation for the h pair (0.100, 0.05) gives $I = 0.199\,999\,2709$ while from the (0.05, 0.025) pair we get $I = 199\,999\,9545$. We now do an h^4 extrapolation (i.e. $16:-1$ mixture) to correct for the h^4 term in the error series and get the estimate $0.2 + 1 \times 10^{-10}$ which is very good, particularly since a little rounding error must be present. Using the Euler-Maclaurin series directly gives the result

$$I = 0.198\,336\,2500 + 0.001\,666\,6667 - 0.000\,002\,9167$$

$$= 0.2 \text{ exactly (at } h = 0.1). \tag{14}$$

The higher terms in the power series vanish identically for $y(x) = x^4$ and the residual terms must be negligible. Indeed for any finite polynomial $y(x)$ the residual term is zero, so the Euler-Maclaurin coefficients can be found using x, x^2, etc in trial integrations.

5.5 Change of variable

The evaluation of the integral of $(1 + x^2)^{-1}$ between 0 and ∞ is painfully slow if a constant stripwidth $\Delta x = h$ is used. One device often recommended in treatments of numerical integration is the introduction of the new variable y, defined by $x = e^y$. I find that in some of my calculations (e.g. those in chapter 10) it is an extra nuisance to have to start at $y = -\infty$ in order to get $x = 0$. I often prefer to set $x = e^{Ky} - 1$, so that $y = 0$ corresponds to $x = 0$ and so that the parameter K can be adjusted to control the rate at which the x axis is traversed if the fixed interval $\Delta y = h$ is used in the integration. With this change of variable the integral of $(1 + x^2)^{-1}$ becomes

$$I = \int_0^\infty K(1 + x)(1 + x^2)^{-1}\, dy \tag{15}$$

with $x = e^{Ky} - 1$. I have written the integral in what I call its 'computational form'. If the traditional 'y variable everywhere' form is directly translated into a program statement it will mention EXP(K * Y) twice. What we need for program economy is to work out x once and then work out the integrand in x-language. The dy reminds us to use an interval h in y, not x. If we choose h values so that, say, $y = 5$ always lies at the end of a strip, then at $y = 5$ we can change K, say from 1 to 2, to speed up the integration beyond $y = 5$. If the same prescription is followed for each h then the series for the error is still one in the quantity h^2, so Romberg analysis may be used. This is still so if K is made a continuously varying function of y, e.g. $K(y) = 1 + y$, so that $y = 0$ still gives $x = 0$. We simply change the integrand by putting $[K(y) + yK'(y)]$ in place of K (and, of course, writing $x = e^{K(y)y} - 1$ under the equation, so that we remember to change correctly all the rows in the program). For many microcomputers the use of a continuous $K(y)$ will be quicker, since the operator can explicitly write in $[K(y) + yK'(y)]$ as a function; it will be $1 + 2y$, for example, if $K(y)$ is $1 + y$. To make K jump from 1 to 2 at $x = 5$ would require either many (slow) logical tests such as (for the midpoint rule)

IF Y > 5 THEN LET K = 2

or it would involve a stop at $y = 5$ with the operator inserting $K = 2$ manually and then using CONT to continue the integration. It would also involve using only h values which fit the interval 0 to 5 exactly.

As an example I have evaluated the integral (15) with $K = 2$ using the very accurate TI-58 calculator and the midpoint rule. To get the integral 'to infinity' I just let the calculator run until the digits had reached a limit, and then waited a while to see if the three guard digits would accumulate to raise the last display digit. The results can be set out in a table as shown below

h	I	$4:-1$	$16:-1$
0.0125	1.570 822 373		
0.025	1.570 900 577	0796 305	
0.050	1.571 214 336	0795 991	6326
0.100	1.572 484 739	0790 868	6332
0.200	1.577 804 723	0711 4110	6165

Any obvious digits have been omitted in each column; this is also done when compiling tables of differences for a function (§4.2). To eliminate each error term in turn from the Euler-Maclaurin series we take weighted averages of numbers in the preceding column, as shown by the column headings. By looking at the series the reader should be able to see that, if h is multiplied by a factor f at each successive integration, then the weighting factors used should be $f^2:-1$, $f^4:-1$, and so on. I used $f=2$, the original Romberg procedure, in my example. The results suggest that the h^2 and h^4 error terms have opposite sign, and show that the operator has to use some judgment. Only by taking the $h=0.0125$ term do we get the 'true' limit of $1.570\,796\,326$, which differs from the analytical value of $\pi/2$ by -1 in the last digit. Romberg's work assures us that we will get to the true limit (barring rounding errors) if we keep on reducing h.

Romberg's procedure has many similarities to the use of Aitken or Padé methods to accelerate the convergence of a sequence; indeed the problem can be tackled by using Padé approximant methods [3, 4] and also by fitting the $I(h)$ values to a rational fraction in h^2 (instead of a power series) to give improved convergence [5]. One obvious point is that Aitken's procedure uses three I values, because it needs to find the apparent common ratio hidden in the series (exercise 3.4). If we know that the form of $I(h)$ is $I(0) + Ah^2 + Bh^4 + \ldots$ then we can make do with two terms at a time, putting in our *a priori* common ratios f^2, f^4, etc in the Romberg table. Applying a repeated Aitken process across the table opposite gives the estimate $I = 1.570\,796\,315$, which would need one more integral value in the table to improve it further. Alternatively, we could think of the tabulated $I(h)$ values as the sums of a convergent series if we read up the table. The 'simulating series' for any sequence $S_0, S_1, S_2 \ldots$ is

$$S_0 + (S_1 - S_0)\lambda + (S_2 - S_1)\lambda^2 + \ldots \tag{16}$$

with λ set equal to 1 when the sum has been evaluated. This series could be put into the Wynn algorithm program (§6.4); however, it is easier to modify the program so that the S_n values (i.e. the $I(h)$) can be put in directly by the operator. To reduce rounding errors we can omit the 1.57 from all the I values and use integer values; thus we use $822\,373$ for $1.570\,822\,373$, $2484\,739$ for $1.572\,484\,739$, and so on. The reverse translation can be made by eye when the approximants are displayed. The reader may verify that the $[2, 2]$ approximant is $796\,319$, giving an error of only -8 in the last digit.

Historical note

The name Romberg integration [6] is usually given to the methods described above. The name of Richardson [7] is usually associated with such methods when applied in other areas. Neither of them thought of it first; the only safe references seem to be Newton and Archimedes!

The detail of an integration program will depend on the integration method employed. I give below a typical flowchart for a trapezoidal rule Romberg integration using stripwidths h, $2h$ and $4h$. It incorporates several devices for saving time and improving accuracy which the reader should find useful in any integration program. (See also exercise 7.)

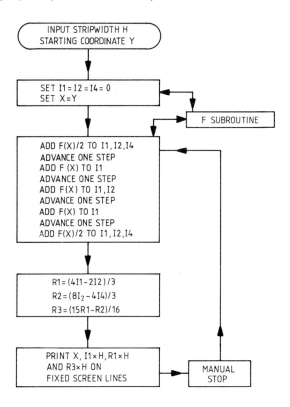

5.6 Numerical differentiation

Expanding the quantity $D(x, h) = \frac{1}{2}h^{-1}[y(x + h) - y(x - h)]$ as a Taylor series shows that it formally equals the derivative Dy at x plus a series in h^2. The Richardson analysis can thus be applied, using the quantities $D(x, h)$, for fixed x, instead of the $I(h)$. In §9.4 I use this procedure for the interesting task of finding expectation values such as $\langle\psi|x^N|\psi\rangle$ without explicit knowledge of the eigenfunction ψ. In §3.3 I have used a crude finite-difference form of Dy in a Newton–Raphson program. Although many textbooks on numerical analysis warn against the pitfalls of numerical differentiation, I find that Richardson analyses work well for the fairly well-behaved functions appearing in much of applied quantum mechanics. I was delighted to make the belated discovery of a short paper by Rutishauser [8] which contains a similar viewpoint. The traditional problem with numerical differentiation is that to get a good Dy value directly we need $D(x, h)$ for $h \to 0$, but will lose accuracy at small h because $y(x + h)$ and $y(x - h)$ will agree in their first few digits. For example, $(1.234\,234 - 1.234\,100) = 0.000\,134$ gives a result with around 1 in 10^2 accuracy, while the original numerical values had accuracy of 1 in 10^6. The Richardson analysis, often called 'the deferred approach to the limit', enables us to extrapolate into the $h \to 0$ limit region without encountering this problem. From the viewpoint of integration, we can use the midpoint rule formula to get

$$\int_{x-h}^{x+h} Dy(x)\, dx = y(x + h) - y(x - h) \tag{17}$$

$$= 2h\, Dy(x) + \tfrac{1}{6}h^2\, [D^2 y]_{x-h}^{x+h} - \ldots \tag{18}$$

$$= 2h\, Dy(x) + \tfrac{1}{3}h^3\, D^3 y(x) + \ldots \tag{19}$$

We simply reproduce the Taylor series result on the right, of course, but this calculation illustrates that numerical differentiation implicitly uses the *known* integral to get the *unknown* midpoint rule sum $h\, Dy(x)$ for a single strip integration. This interpretation makes it clear that taking h too large can give a residual error effect just as it can in integration. The sad fact is that the convergence attainable by using a sequence such as $h, h/2, h/4, \ldots$ in the Richardson process (neglecting rounding errors) is in general not attainable by using, say, the sequence $h/4, h/2, h, 2h, \ldots$.

The 'limit' attained, even if well defined, will contain a residual error of magnitude determined by the smallest stripwidth employed. When the theory of a Richardson process (or any other process) can be based on a Taylor expansion, then it is usually possible to give an upper bound to the error, including the residual error. (In actual computations, of course, there is also a rounding error effect to be considered.) The most simple error estimates arise from the Taylor series with remainder, which says

$$y(x + h) = S(N - 1) + \frac{h^N}{N!}\, D^N f(y + \theta h) \tag{20}$$

where $S(N - 1)$ is the Taylor series sum up to the h^{N-1} term and $0 \leqslant \theta \leqslant 1$. This usually 'feeds through' to give a similar prescription for any other series appearing in a formula: for example if the next term (beyond those already used) in an Euler–Maclaurin series is known then we convert it to give the exact error by the prescription

$$A_N h^{N+1} [D^N y]_A^B \to A_N h^{N+2} (B - A)\, D^{N+1} y(\beta) \tag{21}$$

with $A < \beta < B$. If the integrand y is a polynomial of degree N between A and B this prescription gives the exact size of the error; otherwise the maximum and minimum values of $D^{N+1}y$ must be used to get limits on the error, since β is not known *a priori* and is h dependent, tending to A as h tends to zero. The rough rule of thumb for calculations with asymptotic series is that the error is less than the size of the first neglected term. For series of Stieltjes (§6.2) this can be proved rigorously.

5.7   Endpoint singularities

A change of the variable of integration is often tried in analytic integration, and an example of its use in numerical integration was given earlier. There are several other cases of relevance in physics where such a procedure is useful. As a mathematical example I cite the change of variable

$$x = y(1-y)^{-1}; \quad dx = (1-y)^{-2} \, dy \tag{22}$$

which converts an infinite integral with $0 \leqslant x \leqslant \infty$ into a finite one with $0 \leqslant y \leqslant 1$. Many papers on integration theory treat the standard region 0 to 1 (or -1 to 1) and assume that a standardising transformation can be done to transform any integral into one over the interval. By changing the variable we change the integrand and so may change the nature of the leading term which contributes to the error in the Euler–Maclaurin series. As an example the change $x = y^6$ in our test integral of $(1 + x^2)^{-1}$ will convert the integrand to $6y^5(1 + y^{12})^{-1}$, with a zero value for Dy_0 and $D^3 y_0$. A midpoint integration with $\Delta y = h = 0.1$ gives a value of $1.570\,796\,643$, while the result at $h = 0.05$ is $1.570\,796\,327$. Sag and Szekeres [9] have discussed other ways of rendering the leading h^n error terms zero. For our simple example the theoretical error law should start with terms in h^6 and h^{18}. This leads to a predicted value with error -5×10^{-9}, which I suspect is due to residual (not rounding) error, since a computation at $h = 0.025$ gives $I = 1.570\,796\,326$.

In the case of integrands with endpoint singularities both theory and computation show that the error series can contain functions of h other than h^2, h^4, etc. I consider first the case of a classical oscillating particle with the energy function

$$E = \tfrac{1}{2}mv^2 + Kx^N \tag{23}$$

with $N = 2, 4, 6$, etc. The restoring force is $-NKx^{N-1}$, so the oscillations are not simple harmonic unless $N = 2$. If the particle starts from rest at $x = X$, then the speed at any instant is related to the position, since conservation of energy means that

$$\tfrac{1}{2}mv^2 = K(X^N - x^N).$$ (24)

The time for one oscillation will be four times that for the first quarter of an oscillation; using $v = dx/dt$ we find

$$4 \int_X^0 dt = 4 \sqrt{\frac{m}{2K}} \int_0^X (X^N - x^N)^{-1/2} dx$$ (25)

$$= 4X^M \sqrt{\frac{m}{2K}} \int_0^1 (1 - y^N)^{-1/2} dy$$ (26)

where we set $x = Xy$ and $M = 1 - (N/2)$. The factor outside the integral can be found from dimensional analysis (except for the purely numerical factor) and the integral need only be evaluated once. Only for $N = 2$ is the periodic time amplitude-independent. The integrand has an infinity at $y = 1$, but the midpoint rule will still work, since it does not use the value of the integrand at $y = 1$. With $N = 4$ I found the estimates $I(0.1) = 1.216\,116\,85$, $I(0.05) = 1.243\,655\,913$, $I(0.025) = 1.263\,297\,876$. The ratio of the successive differences in the I values is 1.40. If the error varies as h^k then this ratio is 2^{-k} when h halves at each step. The result suggests that an $h^{1/2}$ law is involved and theory also predicts this [4, 10]. It would require a very small h to produce an I value correct to several digits. Even a Romberg analysis would be tedious, since it turns out that the error series has terms in $h^{1/2}, h^{3/2}, h^2, h^{5/2}$, etc. The simple way to deal with the problem is to change the variable [11] and get back to a regular Romberg analysis with a series in h^2. To do this we write (for $N = 4$)

$$(1 - y^4) = (1 - y)(1 + y + y^2 + y^3)$$

with a similar procedure for other N. We then set $y = 1 - x^2$ and find that

$$\int_0^1 (1 - y^4)^{-1/2} dy = 2 \int_0^1 (1 + y + y^2 + y^3)^{-1/2} dx$$ (27)

with $y = 1 - x^2$. The geometric series in y here has the useful property of being capable of forwards or backwards nested multiplication: we simply do the operation 'times y, plus 1' the requisite number of times. The midpoint rule now gives $I(0.1) = 1.310\,195\,26$, $I(0.05) = 1.310\,820\,441$, and $I(0.025) = 1.310\,976\,694$, with a difference ratio of 4.001. The extrapolated value is $I = 1.311\,028\,774$.

A similar change of variable is useful in connection with the WKB or semi-classical approximation. The Schrödinger equation for the anharmonic oscillator treated above is

$$\frac{-\hbar^2}{2m} D^2 \psi + Kx^4 \psi = \epsilon\psi.$$ (28)

By dimensional analysis or change of scale (§2.5) we find that the bound state energies take the form

$$\epsilon_n = \frac{\hbar^2}{2m} \left(\frac{2mK}{\hbar^2}\right)^{1/3} E_n \tag{29}$$

where the E_n are the energies for the Schrödinger equation

$$-D^2\psi + x^4\psi = E\psi. \tag{30}$$

For the case of the potential x^2, the harmonic oscillator, the E_n can be calculated explicitly (see exercise 7.1 or any quantum mechanics textbook). They take the form $E_n = (2n + 1)$ for integer n, with $n = 0$ for the ground state. The WKB method studies the energy-dependent integral

$$I(E) = \int_{X_1}^{X_2} [E - V(x)]^{1/2}\, dx \tag{31}$$

where $V(x)$ is the potential function and X_1 and X_2 are the classical turning points at which $V(X_1) = E$, $V(X_2) = E$, respectively. For the classical x^4 oscillator treated previously we have $X_1 = -X_2 = X$, the initial amplitude. The integral $I(E)$ can be worked out analytically for the case $V(x) = x^2$ and $E_n = 2n + 1$; the result is

$$I(E_n) = (n + \tfrac{1}{2})\pi. \tag{32}$$

The first-order WKB approximation gives the result that (32) also holds for any single-minimum $V(x)$. Titchmarsh [12] has shown that the multiplier of π should be $(n + \tfrac{1}{2})$ plus a term of order n^{-1} and some workers [13] have estimated the size of the correction term for perturbed oscillator problems. However, the simple formula gives good energy estimates even for moderately excited states $(n \geqslant 5)$. To apply it to the case $V = x^4$ we set $E = X^4$, $x = Xy$, and obtain the WKB formula in the form

$$(n + \tfrac{1}{2})\pi = 2E^{3/4} \int_0^1 (1 - y^4)^{1/2}\, dy. \tag{33}$$

The integral is one with a singularity at the endpoint $y = 1$, but the infinity is in the derivative of the integrand and not the integrand. Using the midpoint rule gives an error series with terms $h^{3/2}$, $h^{5/2}$, etc and we can 'regularise' the integration by the same change of variable used before, i.e. $y = 1 - x^2$. This gives

$$I = 2 \int_0^1 x^2(1 + y + y^2 + y^3)^{1/2}\, dx \tag{34}$$

with $y = 1 - x^2$. The integrand has derivatives 0 and 2 at $x = 0$ and $x = 1$, so the Euler–Maclaurin series becomes

$$I = I(\text{midpoint}) + \tfrac{1}{12}h^2 + \dots. \tag{35}$$

I find that this leads to the estimates 0.874 019 3684 at $h = 0.1$, 0.874 019 1877 at $h = 0.05$, and 0.874 019 1849 at $h = 0.025$. Dividing $I(0.1) - I(0.05)$ by $I(0.05) - I(0.025)$ gives the difference ratio 64.5 and shows that the next term in the Euler–Maclaurin series is the h^6 one, and the extrapolated value becomes $I = 0.874 019 1848$. This gives the first-order WKB result

$$E = 2.185\,069\,30\,(n + \tfrac{1}{2})^{4/3} \tag{36}$$

which can be used to give a good starting energy for input into the very accurate methods of Chapter 7. At $n = 10$ the WKB energy is 99.968 per cent of the exact energy and at $n = 20$ it is 99.992 per cent of the exact energy.

5.8   Multiple integrals

Double integrals over the radial coordinates of two electrons, with integrands such as $r_1^m r_2^n \exp(-\alpha r_1 - \beta r_2)$, appear in atomic theory and are discussed in Appendix 1, which gives a formula for their analytic calculation when m and n are integers. Integrals involving $\exp(-\alpha r)$ and $\exp(-\alpha r^2)$ factors can be converted into one another by a change of variable. The $\exp(-\alpha r^2)$ factor, giving so-called Gaussian orbitals, is favoured in some parts of quantum chemistry, particularly where multi-centre integrals are involved, since a product of Gaussian functions on two different origins can be expressed in terms of Gaussians centred on some third point. However, in simple atomic problems it has been found that many Gaussian orbitals are needed to represent a typical atomic orbital; functions with $\exp(-\alpha r)$ factors give a more compact basis set for atomic problems.

If we consider a double integral such as

$$\int_0^2 \int_0^1 F(x, y) \, dx \, dy \tag{37}$$

and treat it as a repeated integral, then we first do the x integral for a sequence of increasing y, to get a function $G(y)$. We then integrate $G(y)$ to get the final answer. If stripwidths h_1 and h_2 are used in the x and y directions, respectively, together with the midpoint rule, then the error series will in general be a sum of terms of type $h_1^a h_2^b$, where a and b are even integers. However, we can fix the *ratio* h_2/h_1, so that $h_1 = 2^k h$ and $h_2 = r2^k h$, with r constant. The error series then becomes one in h^2, h^4, etc, to which the usual Romberg analysis applies. I used the following program on the Pet to get midpoint estimates of the integral

$$I = \int_0^2 \int_0^1 \exp\left(-x - y\right) \mathrm{d}x\ \mathrm{d}y \tag{38}$$

which has the value 0.546 572 344 to nine figures.

```
 5  INPUT NS
10  H1 = 1/NS : H2 = 2/NS : S = 0
15  DEF FNA (X) = EXP (−X − Y)
20  FOR M = 1 TO NS : Y = (M − 0.5) * H2
20  FOR N = 1 TO NS : X = (N − 0.5) * H1
30  S = S + FNA(X)
40  NEXT N : NEXT M
50  I = S * H1 * H2 : PRINT I
```

The user-defined function facility is used on line 15. The number of strips NS is the same for both x and y directions, although all that is strictly required is that $NS(x)$ and $NS(y)$ should be in a fixed ratio throughout the calculation. I obtained the following Romberg table:

NS	I		
5	0.542 041 607		
10	0.545 435 158	656 6342	723 41
20	0.546 287 764	657 1966	

To evaluate integrals in many dimensions requires a large number of function evaluations and Monte-Carlo methods are sometimes employed [14]. While such methods can be used on a microcomputer giving random numbers, I find that for the simple integrals treated here the Monte-Carlo approach is very slow, since the accuracy typically varies as $N^{-1/2}$ where N is the number of trials (see Appendix 3). Gaussian methods (exercise 10), in which the weights and the sample points are all optimised, can give very accurate results, but require storage of the (irrational) weights and x values to be employed. I find that many of the textbook tables of such quantities are not given to a sufficient number of figures for use on modern microcomputers. As an interesting example of a kind of 'Gaussian differentiation', Ash and Jones [15] have pointed out that to get the value of the slope $Dy(0)$ from three function evaluations the optimum sample points to use are $x = \beta$ and $\beta \pm 1$, with $\beta = (1/\sqrt{3})$. Ralston and Rabinowitz [16] give a lengthy comparison of Gaussian and Romberg approaches and discuss how the error may be estimated in some cases.

I used the simple WKB approximation in §5.7 to give an integral with a singularity. Readers interested in WKB-like methods may like to look at a recent work [17] in which the integral condition (32) is rendered exact, at the expense of using an integrand which is obtained by solving an auxiliary differential equation. The idea of this approach is due to Milne, but it has been revived recently by various authors.

Exercises

1. The sum to infinity of the powers n^{-2} of the integers is $\pi^2/6$. The sum taken up to and including $n = N$ will be less than $\pi^2/6$ by an amount $\epsilon(N)$ which tends to zero as $N \to \infty$. Using the theory of midpoint integration, find the leading terms of a series in N^{-1} for $\epsilon(N)$.

2. Treat the integral of $(1 - y^2)^{1/2}$ between 0 and 1 by the ordinary midpoint integration theory and show that terms in half-integer powers arise in the error series.

3. Consider the evaluation of the integral of the function e^{inx} from 0 to 2π, n being an integer. What does the Euler–Maclaurin series for the midpoint rule suggest for the error? Is this correct?

4. Supposing that the term $(n + \frac{1}{2})$ in the WKB formula should have a correction term T_n added to it, find T_n from the results quoted in the text for $n = 10$ and $n = 20$, and then estimate the energy for $n = 5$.

5. Consider how to change variables to make the following integrals with singularities into integrals without singularities. (All of them have been treated by more complicated techniques in the mathematical literature! Simplify, ever simplify!)

(a)
$$\int_0^\infty \frac{e^{-x}\,dx}{(x + 1)x^{1/2}} \tag{39}$$

(b)
$$\int_0^1 x^{1/2}(1 - x)^{1/2}\,dx \tag{40}$$

(c)
$$\int_0^1 t^{-3/4}(1 - t)^{-1/2}h(t)\,dt \tag{41}$$

6. Work out $(1 - z)^{-1}$ at $z = 0.937$ on your computer. Can you see how to get more digits by using an idea similar to that in the square root calculation of exercise 1.2?

7. Consider an integrand without any singularity problem, and suppose that it is to be evaluated using stripwidths $h, 2h, 4h$, etc. Does the trapezoidal rule have any computational advantage over the midpoint rule?

8. Consider the following values of a function. Can you estimate the slope at $x = 1$?

x	f
0.7	4.2001
0.8	4.0120
0.9	3.9512
1.0	4.0000
1.1	4.1492
1.2	4.3947
1.3	4.7355

9. Can the midpoint and trapezoidal rules be more accurate than Simpson's rule at the same stripwidth?

10. Simpson's rule uses the $y(x)$ values at fixed intervals h. To find the exact integral of x^N, with $N = 0$, 1, 2, or 3, it would suffice to use the values of the integrand at three points. To find the integral between 0 and 1, then, an h of $\frac{1}{2}$ would suffice and use of $h = \frac{1}{4}$ would give no improvement. If the sample points are not equidistant, can we achieve similar exact results with fewer than three points?

Solutions

1. From the midpoint rule we have (with $h = 1$)

$$\int_{N-1/2}^{\infty} x^{-2}\, dx = \sum_{N}^{\infty} n^{-2} + \tfrac{1}{12}(N-\tfrac{1}{2})^{-3} \ldots. \tag{42}$$

This yields

$$(N-\tfrac{1}{2})^{-1} = \epsilon(N-1) + \tfrac{1}{12}(N-\tfrac{1}{2})^{-3} + \ldots. \tag{43}$$

This yields

$$\epsilon(N-1) = N^{-1} + \tfrac{1}{2}N^{-2} + \tfrac{1}{6}N^{-3} + O(N^{-5}) \tag{44}$$

with no N^{-4} term. I rediscovered as a schoolboy the fascinating result that the first term in such sums of powers is what you get by integrating, while the next coefficient is always $\frac{1}{2}$. The same applies for sums from 1 to N of positive powers of the integers, e.g. $\Sigma_1^N n = \frac{1}{2}N^2 + \frac{1}{2}N$.

2. The derivative of $(1-y^2)^{1/2}$ is formally $-y(1-y^2)^{-1/2}$. At $y = 1-x$ the function equals $[x(2-x)]^{1/2}$. Splitting the integral into two parts $0 \to (1-h)$ and $(1-h) \to 1$, it is clear that the integral from $(1-h)$ to 1 is

also the integral from 0 to h of $x^{1/2}[2-x]^{1/2}$. This yields a series of terms in $h^{3/2}, h^{5/2}$, etc. The integral from 0 to $(1-h)$ yields an endpoint product $h^2 Df$ with leading term $h^2 h^{-1/2} = h^{3/2}$ at the $(1-h)$ end and a zero at the other end. Going to the higher terms we get no contribution from the $y = 0$ end but a series with terms $h^{3/2}, h^{5/2}, h^{7/2}$, etc from the other end.

3. The integral is of the type appearing in the theory of Fourier series and has the value

$$\int_0^{2\pi} e^{inx}\, dx = 2\pi\delta_{no}. \tag{45}$$

Because the integrand is periodic with period 2π all the Euler-Maclaurin coefficients vanish, which implies that the midpoint sum or the trapezoidal sum give exact results for any h which fits the interval. For example, if the trapezoidal sum is taken with N strips the contribution from the point $x = Mh$ is

$$h \exp[i2\pi nMN^{-1}]. \tag{46}$$

The total trapezoidal sum is zero if n is not equal to 0 or a multiple of N; otherwise it is $Nh = 2\pi$. (Use an Argand diagram to see this.) Thus the trapezoidal rule gives zero error if $N > n$. If the trapezoidal sum is taken for the product $f(x)e^{inx}$, where $f(x)$ has Fourier coefficients A_m, the sum is

$$h \sum_m \sum_M A_m \exp[i2\pi(n+m)MN^{-1}]. \tag{47}$$

The integral of the product would give $2\pi A_{-n}$, but the sum gives a sum of terms A_k, with $k = -n +$ any multiple of N. Using sums over N points gives what is called a discrete Fourier transform in which each coefficient is an infinite sum over the usual Fourier coefficients. Nowadays there are various fast Fourier transform methods for calculating the discrete transform quickly. If there are, say, 20 sample points (i.e. $N = 20$) then we can work out discrete Fourier coefficients with $0 \leqslant n \leqslant 19$. (Going outside this range would duplicate the coefficients already calculated.) To get all the coefficients it looks as though it needs 20×20 terms to be calculated, but in fact by manipulating the exponential factors it is possible to cut down the computing time by a factor of roughly ten. In general the time ratio involves the sum of the factors over the product of the factors; for example if $N = 128 = 2^7$ the product of its factors is 128 but their sum is $7 \times 2 = 14$. Cooley and Tukey [18] give the relevant theory and two recent articles [19, 20] give BASIC programs for working out fast discrete transforms.

4. Taking the n^{-1} term to be βn^{-1} the results at $n = 10$ give

$$(10.5 + T_{10})^{4/3} = 0.999\,68(10.5)^{4/3}. \tag{48}$$

This gives $T_{10} = 0.00252$. The results at $n = 20$ give $T_{20} = 0.00123$. Using the rough estimate $T_n = 0.025n^{-1}$, we find the result $E_5 = 21.23937$. The accurate value is 21.23837 (to five places). The uncorrected formula gives 21.21365.

5. The changes of variable are:

(a) $x = y^2$.

(b) $x = 1 - y^2$, then $y = 1 - z^2$, so that $x = z^2(2 - z^2)$. In the analysis the integrand becomes $4(2 - z^2)^{1/2}z^2(1 - z^2)^2$. In the computation we can go back to $x^{1/2}(1 - x)^{1/2}$ and work out x at each z value. All the analysis does is to tell us the function $x(z)$.

(c) $t = 1 - y^2$, then $y = 1 - z^4$, so that $t = z^4(2 - z^4)$. The integrand in this case becomes $8h(t)(2 - z^4)^{-3/4}$, so that use of the integral in this z, t form is as easy as going back to the original integrand.

6. Suppose we get $(1 - z)^{-1} = 15.873016$. We set

$$(1 - z)^{-1} = 15.87 + h \tag{49}$$

which gives

$$h = (15.87z - 14.87)/(1 - z). \tag{50}$$

Using $z = 0.937$ in this gives (if the computer uses the mantissa-exponent display) $h = 3.0158730\mathrm{E} - 3$, producing the more accurate estimate 15.8730158730. Tricks like this also work on small calculators which do not have exponent notation: the operator simply calculates $15.87z - 14.87$ to be 0.00019 and then uses 19 instead to gain extra digits [21].

7. To start at $x = 0$ and use the midpoint rule needs the values of the integrand at $\frac{1}{2}h, \frac{3}{2}h$, etc. Repeating with twice the stripwidth would use the points $h, 3h$, etc. For the case of the trapezoidal rule the points used on each run would be $0, h, 2h \ldots$ and $0, 2h, 4h \ldots$, so that both integrals, and also those for $3h$ and $4h$, can be worked out on one run if the program is correctly written. If we suppose that working out $y(x)$ is the longest process involved in a step, it follows that working out two integrals at a time (for h and $2h$) uses $\frac{2}{3}$ of the time that would be used for two consecutive runs. If the program also internally combines the two results in a ratio of $4:-1$ to eliminate the h^2 error term then it will effectively be taking a sum of type

$$\tfrac{4}{3}[\tfrac{1}{2}f_0 + f_1 + f_2 \ldots]h - \tfrac{1}{3}[\tfrac{1}{2}f_0 + f_2 + \ldots]2h \tag{51}$$

$$\tfrac{1}{3}[f_0 + 4f_1 + 2f_2 + 4f_3 + \ldots]h \tag{52}$$

i.e. it will be a Simpson's rule program for stripwidth h. One way to get the program to use, say, four h values at the same time is to treat the four integrals as an array I(N). The reader can check that the following nested loop segment will handle the job. I use the notation FNA for the integrand. On a Pet this would represent a user-defined function; on a PC-1211 it could be represented by a user-defined key. In any other case the integrand function can just be written in.

10 FOR Q = 1 TO 4 : C = C + 1 : X = C * H
20 FOR N = 1 TO Q : T = FNA (X)
30 I(N) = I(N) + T : NEXT N
40 PRINT I(1) : NEXT Q

The accurate counting process in line 10 could be replaced by $X = X + H$ on many machines (§2.2). In line 30 the quantity I(N) is a sum of function values, not multiplied by the stripwidth. This procedure saves time and also reduces rounding errors. At a later stage in the program we can have further instructions, such as

100 FOR N = 1 TO 4
110 I(N) = I(N) * H * N : NEXT N

to convert to the values of the integral. Further instructions can form combinations such as $[4I(1) - I(2)]/3$ etc to make up the numbers which would appear in a Romberg analysis. An instruction telling the program to jump to 100 if FNA(X) is less than, say, 10^{-10}, or if X equals the upper integration limit, would also be included if the program is to run on its own. Special contributions from the endpoints (with weighting factor $\frac{1}{2}$) must be included for the trapezoidal rule, although these are not needed if the integrand is zero at both ends of the region of integration. This situation arises for many of the expectation value integrals arising in the theory of the radial Schrödinger equation.

8. We find $D(1, 0.1) = 0.9900$, $D(1, 0.2) = 0.9567$ and $D(1, 0.3) = 0.8923$. Combining the first two in the ratio $4:-1$ gives 1.0011. Combining all three in the ratio $15:-6:1$, as given in the table in §4.3, gives 1.002. The function values were rounded values of $1 + 2x^{-1} + x^3$, which has a slope of 1 at $x = 1$. The crude estimate $Df = h^{-1}[f(x + h) - f(x)]$ in this case would give a 'right slope' three times as big as the 'left slope', illustrating that symmetrical (central-difference) formulae should be used whenever possible. To estimate the slope at $x = 0.7$ involves using only forward-differences, with an error series involving all powers of h, and the longer calculation will be more affected by rounding errors. The Newton–Gregory interpolation formulae (§4.2) give formally equivalent results to the Romberg analysis. For example, the first two differences at 0.9 are $\Delta f = 0.0488$ and $\Delta^2 f = 0.1004$. With origin at 0.9 the interpolating curve is thus

$$f = 3.9512 + x(0.0488) + \tfrac{1}{2}x(x - 1)(0.1004) \qquad (53)$$

if x is measured in units of 0.1. The slope is thus

$$f' = 0.0488 + (x - \tfrac{1}{2})(0.1004) \tag{54}$$

$$f' = 0.099 \tag{55}$$

which becomes 0.9900 when we allow for the x units of 0.1. This agrees with D(1, 0.1), but was somewhat more painful to calculate. However we did get an f' estimate for every x value as well as just for $x = 1$.

9. If an integrand $y(x)$ gives zero for the factor $[Dy]_A^B$ which multiplies h^2 in the Euler-Maclaurin series, then the leading h^4 error term in both the midpoint and trapezoidal rules is smaller than that in Simpson's rule. The ratio is $-7:8:28$, the midpoint rule giving the coefficient of smallest magnitude. Using Simpson's rule is equivalent to combining two trapezoidal results for stripwidths h and $2h$. The error for these two stripwidths is of form $\epsilon_1 = A + B$, $\epsilon_2 = 4A + 16B$. If A is non-zero then it is removed by forming $(4I_1 - I_2)/3$. This leads to an error $-4B$, making things worse if A happens to be zero. Of course, particularly for infinite integrals, we can usually change variables to make this happen, and produce examples to make Simpson's rule look bad, just as integrating x^3 makes it look good!

10. This is a simple example of Gaussian integration, in which the weighting factors and the sampling points are varied to get the best result. This process appeals to physicists who are used to variational methods in quantum and classical mechanics – well, anyway, it appeals conceptually to me! Nevertheless I don't use it much, because it doesn't lead to an easy process of systematic improvement like the Romberg one; however, for a given number of function evaluations the Gaussian method is more accurate.

   If we use the integration rule

$$\int_0^1 y(x)\, dx = \Sigma\, W_j y(x_j) \tag{56}$$

and try to make it give exact results for the integral of x^N, then we require that

$$\Sigma\, W_j x_j^N = (N + 1)^{-1}. \tag{57}$$

Fitting one point to $N = 0$ and $N = 1$ gives the midpoint rule: $W = 1, x = \tfrac{1}{2}$. Using two points we would intuitively expect left-right symmetry, and can try $x_2 = 1 - x_1$, $W_1 = W_2 = \tfrac{1}{2}$. The equations for $N = 0$ and $N = 1$ are obeyed, while that for $N = 2$ gives

$$\tfrac{1}{3} = \tfrac{1}{2}[x^2 + (1 - x)^2] \tag{58}$$

so that

$$0 = 6x^2 - 6x + 1 \tag{59}$$

$$\text{The roots are } x = \tfrac{1}{2} \pm \tfrac{1}{6} \sqrt{3} \tag{60}$$

(Numerically we could get the root by iterating the input-output formula $y = [6(1-x)]^{-1}$. Using $x_0 = 0.5$ gives the result 0.2113248654.) The resulting integration formula also works exactly for $N = 3$, so we can get away with using only two points, although these have irrational x values. A direct check by computer shows that for $N > 3$ the formula gives a low result. For $N = 4$ it gives the error $-\frac{1}{180}$ which would translate into $-\frac{1}{4320}h^4 [D^3 y]$ for an arbitrary integration region, if within each strip of width h we used sample points at the two Gaussian positions.

Notes

1. Y L Luke 1955 *J. Math. Phys.* **34** 298
2. E T Goodwin 1949 *Proc. Camb. Phil. Soc.* **45** 241
3. J S R Chisholm 1974 *Rocky Mtn. J. Maths.* **4** 159
4. D K Kahaner 1972 *Math. Comput.* **26** 689
5. R Bulirsch and J Stoer 1964 *Numer. Math.* **6** 413
6. W Romberg 1955 *K. Norske Vidensk. Selsk. Forhandlinger* **28** No 7
7. L F Richardson and J A Gaunt 1927 *Trans. R. Soc.* A **226** 299
8 H Rutishauser 1963 *Numer. Math.* **5** 48
9. T W Sag and G Szekeres 1964 *Math. Comput.* **18** 245
10. L Fox 1967 *Comput. J.* **10** 87
11. H Telle and U Telle 1981 *J. Mol. Spectrosc.* **85** 248
12. E C Titchmarsh 1962 *Eigenfunction Expansions* vol 1 (Oxford: Oxford University Press) ch 7
13. F T Hioe, D MacMillen and E W Montroll 1976 *J. Math. Phys.* **17** 1320
14. A Ralston and H S Wilf ed *Mathematical Methods for Digital Computers* 1960 vol 1, 1967 vol 2 (New York: J Wiley)
15. J M Ash and R L Jones 1982 *Math. Comput.* **37** 159
16. A Ralston and P Rabinowitz 1978 *A First Course in Numerical Analysis* (Tokyo: McGraw-Hill Kogakusha)
17. H J Korsch and H Laurent 1981 *J. Phys. B: At. Mol. Phys.* **14** 4213
18. J W Cooley and J W Tukey 1965 *Math. Tables and Aids to Computation* **19** 297
19. B Rogers Dec 1980 *Practical Computing* 91
20. W Barbiz Sept 1981 *Practical Computing* 112
21. J P Killingbeck 1981 *The Creative Use of Calculators* (Harmondsworth: Penguin)

6 Padé approximants and all that

6.1 Power series and their uses

The stern conventional wisdom, still taught to some physics students in their mathematics courses, is that if a power series fails to pass every test for convergence (i.e. all the papers in its entrance examination) then we must have no further truck with it. If, however, it does pass, then it is 'one of the family' and all family members must be treated with equal respect. This strict attitude stems from a period in the history of mathematics when absurd results were being produced (even by the great Euler) by means of analysis which made uncritical use of divergent power series [1]. However, there is by now an extensive mathematical literature dealing with the theory of divergent power series and in recent years work in this field has been much stimulated by problems from quantum mechanics. (It is rather ironic that the traditional perturbation theory of physics textbooks produces in many cases precisely the kind of divergent series which in their mathematics lectures students are urged to discard.) Most physicists already know that a power series, even when officially convergent, may converge so slowly that hundreds of terms must be taken before a stable value for the 'sum' is found. If the sum is found by using a computer the rounding errors may be appreciable in such a long calculation, leading to an incorrect limiting value for the sum. Various ways of rearranging the terms in a series can be tried to get quicker convergence for a convergent series (e.g. the Euler transformation of §4.2). If the sequence of sums of the series is treated then the Aitken procedure (solution 3.4) is often effective. The really interesting effects, however, occur if a *divergent* series is treated by such methods, since there sometimes results a 'sum' for the series which on deeper examination turns out to be mathematically meaningful. In the following section I outline my own approach, using simple methods, to a portion of the mathematics of divergent series which appears fairly often in quantum mechanics. It is the theory of Padé approximants and series of Stieltjes. I try to avoid the classical

80

moment problem which physicists often complain about as an encumbrance, and hope that most physicists will gather at least the main features of the subject from my simple approach.

6.2 Series of Stieltjes

I start by first reminding the reader of the equation

$$1 - z + z^2 - z^3 \ldots = (1 + z)^{-1}. \tag{1}$$

From one viewpoint the function $(1 + z)^{-1}$ is the sum to infinity of the geometric progression on the left, the common ratio being $-z$. If $|z| > 1$, however, the series is one of those wicked divergent ones and we are not supposed to have a sum to infinity. From another viewpoint the series is just the Taylor series (around $z = 0$) for the function $(1 + z)^{-1}$. The fact that the series diverges outside the circle of convergence $|z| = 1$ in the complex plane is then related to the singularity in the function at $z = -1$. To be analytic a function $f(z)$ must have a well defined derivative $f'(z)$; if it does the routine theory shows that higher derivatives also exist. A singularity is a point (or line) at which $f(z)$ is not analytic. $(1 + z)^{-1}$ has a singularity called a pole of order 1 at $z = -1$. The function $(1 + z)^{-1}$ looks to be quite innocent for real positive z, but the series diverges at $z = +1.01$, say, because the size of the circle of convergence is set by the singularity at $z = -1$. In the case of the function $(1 + z^2)^{-1}$, which is a frequent example in works on numerical analysis, the divergence of the Taylor series along the real axis for $|z| > 1$ is determined by poles at $z = \pm i$. These are not on the real axis at all, and this illustrates how (nominally) real variable calculations can be affected in ways which require complex variable theory for a full explanation. The series above for $(1 + z)^{-1}$ is what we would get if we tried to find a power series solution about $z = 0$ for the differential equation

$$(1 + z)f' + f = 0 \tag{2}$$

with the initial condition $f(0) = 1$. In this case we would have no hesitation in saying that what the series 'means' is really the function $(1 + z)^{-1}$, which satisfies the equation for all $z \neq -1$. If the equation is written with $A_1 = 1$.

$$\Sigma A_n(z) D^n f = 0 \tag{3}$$

then $A_0(z)$ is, of course, $(1 + z)^{-1}$ again, with a singularity at $z = -1$. The theory of the power series solution of differential equations (due mainly to Frobenius) says that we can get a convergent power series for f throughout a region of z for which all the functions $A_n(z)$ are analytic (and thus have convergent power series expansions). To put it simply, if the coefficient series are OK, so is the solution series. However, as in our example, we can sometimes 'spot the function' and move on to other z values (analytic continuation). Sometimes this can be done by acting *on the series* with appropriate mathematical operations, and this is where Padé approximants come into the story.

A *series of Stieltjes* is one which arises by formally expanding as a power series in z the quantity

$$F(z) = \int_0^\infty \frac{\psi^2(t)}{(1+zt)} \, dt. \tag{4}$$

The official texts on the theory use $d\phi(t)$ where ϕ is a positive measure function, but I use $\psi^2(t) \, dt$ (without changing the conclusions) so that ψ is analogous to a quantum mechanical wavefunction and $F(z)$ looks like a quantum mechanical expectation value;

$$F(z) = \langle \psi | (1+zt)^{-1} | \psi \rangle \tag{5}$$

$F(z)$ is simply a sum of an infinite number of geometric series, each with its own common ratio. It is clear that contributions from the region $zt > 1$ will give divergent geometric series and so we may expect that the series for $F(z)$ will be divergent in general. By expanding $(1+zt)^{-1}$ and doing the integral term by term we find

$$F(z) = \Sigma \, \mu_n (-z)^n \tag{6}$$

where the nth moment μ_n is the integral

$$\mu_n = \int_0^\infty t^n \psi^2(t) \, dt \equiv \langle \psi | t^n | \psi \rangle. \tag{7}$$

The series for $F(z)$ will usually diverge, even if $F(z)$ is a finite quantity which can be calculated accurately by numerical integration. As an example which can be handled analytically we can take the case $\psi^2(t) = e^{-t}$. The μ_n integrals then give the value $n!$ (Appendix 1), and the series is the Euler series,

$$F(z) = 1 - z + 2!z^2 - 3!z^3 \ldots . \tag{8}$$

Erdelyi [2] gives a detailed discussion of this series and how it can be used to estimate the value of $F(z)$, although he does not deal with Padé approximant methods. Since I am trying to emphasise the computational aspects of various mathematical procedures I will quote some numbers presently. If the coefficient of $(-z)^n$ were $(n!)^{-1}$ we would have the series for e^{-z}, which converges for all z, although a direct computer summation doesn't give very good results for large z values. (Interestingly enough the problem of finding rational approximants for e^z was one of the ones studied by Padé when he established the theory now associated with his name.) With coefficients $n!$ the Euler series diverges for all z. However, it *seems* to be converging at first, and only 'blows up' into violent increasing oscillations after we reach the term of smallest modulus. If this term is the Nth then we must have

$$N!z^N \cong (N+1)!z^{N+1} \qquad (9)$$

from which it follows that $(N+1) \cong z^{-1}$. For $z = 0.1$ we seem to get convergence up to $N = 10$; we will also have the smallest gap between successive estimates of the sum if we look at terms around the smallest one. In this case with $z = 0.1$ the sums are as follows, with S_N denoting the sum up to and including the z^N term:

N	S_N	A_N
6	0.915 92	
7	0.915 416	
8	0.915 8192	0.915 640 00
9	0.915 456 32	0.915 628 21
10	0.915 8192	0.915 637 76
11	0.915 420 032	0.915 629 12

The value of $F(z)$ using numerical integration is 0.915 633 33 at $z = 0.1$, and taking the sum to the smallest term gives a good estimate. In fact the theory given below shows that S_N and S_{N+1} must straddle the exact value, so we would conclude that $F(z) = 0.915\,64 \pm 0.000\,18$ by using the sums of the series, even though the series diverges if we keep on going. The study of the so-called asymptotic series, of which the Euler series is an example, has a long history and several books set out the basic theory in a manner suitable for physicists [2, 3]. The point which makes the subject really fascinating is that one can go beyond that classical theory based on the S_N and get even better results, while still using only the knowledge of the S_N values. For example, in solution 3.4 I described the idea of the Aitken summation procedure, which estimates the limit of a sequence or the sum of a series by proceeding as though a geometric series were involved. We actually know that our example is a series of Stieltjes, with an infinite number of geometric series embedded in it, but can at least try out the Aitken method, using the formula

$$A_N = [S_N S_{N-2} - S_{N-1}^2]/[S_N + S_{N-2} - 2S_{N-1}]. \qquad (10)$$

To avoid rounding error it is easy to drop the common leading digits 915 and set, for example, $0.915\,416 \rightarrow 0.416$, adding back the leading digits after using the formula for A_N. The A_N results look better than the S_N ones and from A_{10} and A_{11} we conclude that $F(z) = 0.915\,633\,44 \pm 4.32 \times 10^{-6}$. (The basic theory shows that the A_N straddle the correct result just as the S_N do.) If we are not inclined to miss out on a good thing we can try the Aitken procedure again on the A_Ns. If we proceed as before we get $F(z) = 0.915\,633\,35 \pm 1.3 \times 10^{-7}$; as far as I know there is not an official theorem that we *must* have the straddling property in this 'double Aitken' procedure as applied to all series of Stieltjes, but there *is* such a theorem for the Padé approximants which I discuss below.

One useful result in the theory of series of Stieltjes is the following one; if S_N denotes the sum of the $F(z)$ series up to and including the z^N term then we find by studying the integral for $F(z)$ that

$$F(z) = S_N(z) + (-z)^{N+1} \int t^{N+1} \psi^2(t)(1 + zt)^{-1} \, dt. \tag{11}$$

If z is real and positive (or a complex number with a positive real part) then the last term has a modulus less than that of the $(N + 1)$th term of the $F(z)$ series, and for real positive z we get the straddling property: $F(z)$ is between S_N and S_{N+1}. The equation above enables us to establish what I call the '*pairing property*' between approximants to $F(z)$; if we find a rule for getting a lower bound to the value of the integral on the right we immediately get an upper or lower bound to $F(z)$, depending on whether N is even or odd. This *pairing principle* enables us to generate many lower and upper bounds starting from a limited number of special lower bounds.

6.3   Padé approximants

Consider the rational fraction P/Q where

$$P = \sum_0^M P_n z^n \tag{12}$$

and

$$Q = \sum_0^N Q_n z^n \qquad (Q_0 = 1) \tag{13}$$

The ratio P/Q can be formally expanded as a power series in z. If this expansion fits to the series expansion of some function $f(z)$ up to the z^{M+N} term, then P/Q is called the $[M/N]$ Padé approximant to $f(z)$. (Some authors use the notation $[N|M]$ for this, but the sloping / in my symbol means divide, making the meaning clear.) To find P and Q we can set $P = fQ$ and compare coefficients, as I illustrate for $[1/2]$ in §6.4. The Padé approximants can, of course, be worked out for any power series $f(z)$, whether or not it is a series of Stieltjes arising from an integral. Indeed in the recent history of quantum mechanics it has often happened that an empirically successful Padé analysis of a perturbation series has preceded a formal proof that it actually 'represents' a function of Stieltjes type. The discovery that the methods work numerically for some non-Stieltjes functions has stimulated a search for the most general class of functions for which the methods can be rigorously *proved* to work. The $[M/N]$ Padé approximant has the simple property that it can be found using only the terms in the series $f(z)$. If $f(z)$ also happens to be a series of Stieltjes then the approximants lead to increasingly accurate upper and lower bounds to the value of $f(z)$ as we increase the order of the approximants, i.e. use information from more and more terms of the series (despite the divergent nature of the series).

One way to approach the calculation of the integral $F(z)$ is to regard it as the expectation value of the inverse $(1 + zt)^{-1}$, as in equation (5). In §3.4 I gave a simple iterative prescription for the calculation of an inverse. When applied to the $F(z)$ problem it leads to the conclusion that the quantity

$$I(\phi) = 2\langle \phi | \psi \rangle - \langle \phi | (1 + zt) | \phi \rangle \tag{14}$$

with ϕ an arbitrary trial function, gives a lower bound to $F(z)$ if z is real and positive. (The relevant details are given in solution 3.) To get an *exact* $F(z)$ value would require us to use $\phi = (1 + zt)^{-1}\psi$, which is possible if we do a numerical calculation. However, the trick is to produce an estimate for $F(z)$ which involves only the *coefficients* in the formal $F(z)$ series, equation (6). To do this we take as the trial function ϕ a sum of terms of form $A(n)(zt)^n$ and vary the coefficients $A(n)$ to make $I(\phi)$ a maximum. The simple choice $\phi = (1 - zt + z^2 t^2)\psi$, for example, gives a ϕ with an error of order z^3. This produces an error of order z^6 in the $F(z)$ estimate and actually gives the $[5/0]$ Padé approximant as the lower bound to $F(z)$. However, using $\phi = [A(0) + A(1)zt + A(2)z^2 t^2]\psi$ and varying the three coefficients to maximise $I(\phi)$ *must* give a better lower bound to $F(z)$. In fact it gives the $[2/3]$ Padé approximant to the series. A similar approach using terms up to $A(N)(zt)^N$ gives the $[N/N + 1]$ approximant *for any series*. For a series of Stieltjes the result is also a *lower bound* to the $F(z)$ value. (See also Appendix 3.)

For the moment I use the brief notation $F(N, z)$ for the integral which differs from $F(z)$ only by having an extra factor t^N in the numerator. The pairing principle formula becomes

$$F(z) = S_N(z) + (-z)^{N+1}F(N + 1, z) \tag{15}$$

where S_N is the sum of the $F(z)$ series up to z^N terms. Suppose that we have the $[0/1]$ approximant to $F(N + 1, z)$. Then we have a lower bound to $F(N + 1, z)$ and get an upper or lower bound to $F(z)$, depending on whether N is even or odd, respectively (assuming throughout that z is real and positive). However, if we express $F(z)$ as a fraction with $Q(z)$ as the denominator, where our starting $[0/1]$ approximant is P/Q, it is clear that the result is the $[N+1/1]$ approximant to the $F(z)$ series, since approximants of given order are unique. The conclusion is that the $[N/1]$ approximant to the $F(z)$ series gives a lower bound to the integral $F(z)$ when N is even and an upper bound when N is odd. More briefly, $(-1)^N [N/1]$ gives a lower bound to $F(z)$. By repeating the argument, starting from the $[1/2]$ lower bound to $F(N, z)$ we conclude that $(-1)^{N+1} [N/2]$ gives a lower bound to $F(z)$, with $N \geqslant 1$. Similarly, for $N \geqslant 2$, $(-1)^N [N/3]$ gives a lower bound, and so on through the whole set of Padé approximants of $[M/N]$ type with $M \geqslant N - 1$. I chose this approach because it makes an infinite set of results arise out of the pairing principle plus a set of special case results, $[0/1]$, $[1/2]$, etc. This appeals to my sense of economy. Further, we only need to be able to compute, say, a $[1/2]$ approximant in order to get an $[N/2]$ approximant by adding on the appropriate S_N. This would avoid evaluating the ratio of two lengthy polynomials and could be done by having a fixed subroutine to evaluate $[1/2]$, with the main program selecting which of the coefficients of the $F(z)$ series to feed to it. However, it turns out that there is a smart recursive algorithm which will get us the $[M/N]$ even more simply (§6.4).

My simple sketch above omits some important details. For example, it follows from the traditional theory that $S_4 = [4/0]$ gives an upper bound, and from our argument above $[3/1]$ and $[2/2]$ will be increasingly better upper bounds using only the same set of terms in the series. Also a lower bound sequence comes from $[5/0]$, $[4/1]$, $[3/2]$ and $[2/3]$. The problem is: do the sequences $[N/N]$ and $[N-1/N]$ converge as N increases and do they converge to the same limit? One sufficient condition for the answers to be yes is that the sum of the terms $\mu_n^{-1/(2n+1)}$ shall diverge. Recently it has been shown analytically [4] that the ground state perturbation series for the energy $E(\lambda)$ for the perturbed oscillator Schrödinger equation

$$-D^2 \psi + x^2 \psi + \lambda x^N \psi = E(\lambda)\psi \tag{16}$$

does give upper and lower Padé approximant bounds which converge and meet for $N = 2$, 4 and 6. However, for $N = 8$ the upper and lower bound sequences give limits which in principle need not agree. Whether the 'gap' is large enough to be of numerical importance is not clear; as far as I know nobody has yet done sufficiently accurate computations for the (viciously divergent) series to exhibit the gap numerically.

6.4   Computing Padé approximants

Consider the task of finding a $[1/2]$ Padé approximant for the $F(z)$ series. We set $F(z)$ equal to the ratio N/D, with

$$N = N_0 + N_1 z \tag{17}$$

$$D = D_0 + D_1 z + D_2 z^2 \tag{18}$$

and follow the usual convention $D_0 = 1$. Since $N = DF$ we find

$$N_0 + N_1 z = (D_0 + D_1 z + D_2 z^2)(\mu_0 - \mu_1 z + \ldots). \tag{19}$$

Comparing coefficients gives

$$N_0 = \mu_0 D_0$$
$$N_1 = -\mu_1 D_0 + \mu_0 D_1$$
$$0 = \mu_2 D_0 - \mu_1 D_1 + \mu_0 D_2 \tag{20}$$
$$0 = -\mu_3 D_0 + \mu_2 D_1 - \mu_1 D_2 .$$

With $D_0 = 1$ we find $N_0 = \mu_0$, and also convert the problem of finding the D_n into a 3×3 matrix problem. The solution is

$$N_1 = \mu_0 D_1 - \mu_1$$
$$D_2 = k(\mu_1 \mu_3 - \mu_2 \mu_2) \tag{21}$$
$$D_1 = k(\mu_0 \mu_3 - \mu_1 \mu_2)$$

with

$$k = (\mu_0 \mu_2 - \mu_1 \mu_1)^{-1}. \tag{22}$$

My intention here is to illustrate that the solution for the Ns and Ds can be obtained in principle by matrix manipulations, and the formal result for Padé approximants given in most textbooks represents them as a ratio of two determinants. If we ignore the exceptional cases where the matrices concerned are singular we get a single answer, i.e. a Padé approximant of a given order is unique. If by any means we get a polynomial N of degree A and a polynomial D of degree B, such that N/D agrees with a series $F(z)$ just up to the z^{A+B} term, then N/D is *the* $[A/B]$ approximant to $F(z)$. (Exercise 7 gives an example.)

To find the $[5/2]$ approximant, say, it might seem that we have to work out a ratio of two large determinants, giving a danger of rounding errors. However, by using the pairing principle idea we can simply set

$$F(z) = \mu_0 + \ldots + z^4 [\mu_4 - \mu_5 z + \mu_6 z^2 - \mu_7 z^3] + \ldots. \tag{23}$$

If we replace the term in the square bracket by its $[1/2]$ Padé approximant we get the $[5/2]$ approximant to $F(z)$. To see this we make $F(z)$ into a fraction with $[1 + D_1 z + D_2 z^2]$ as the denominator. The result is a ratio of a fifth-order polynomial to a second-order one, and by the uniqueness theorem this must be $[5/2]$.

In 1956 Wynn [5] published a brief but remarkable paper in which he showed that a simple repetitive algorithm would produce the values of the various Padé approximants for a given z value. The clever part of his work was, of course, the *analysis* which proved this by manipulating the traditional formulae involving determinants. However the resulting algorithm is as follows. Set out the sums S_0, S_1, etc, as a column and interlace them with a column of zeros to the left, as shown below

$$
\begin{array}{cccccc}
 & S_0 & & \bullet & & \\
0 & & \times & & \bullet & \\
 & S_1 & & \otimes & & \bullet \\
0 & & \times & & \times & \\
 & S_2 & & \times & & \\
0 & & \times & & & \\
 & S_3 & & & &
\end{array}
$$

Now compute the numbers in the crossed positions by using the 'lozenge algorithm'

$$D : \quad D = A + (C - B)^{-1}. \tag{24}$$

The numbers in the S column (column 2) and in the other *even* columns are Padé approximant values. For example, the circled cross clearly uses information from S_0, S_1 and S_2 and is the $[1/1]$ approximant, while that below it is $[2/1]$. Wynn's algorithm as usually quoted gives approximants $[M/N]$ with $M \geqslant N$. However, it seemed to me that Wynn's algebra applied to more general cases and I tried putting into the algorithm a value for $[0/1]$ i.e. $\mu_0^2(\mu_0 + z\mu_1)^{-1}$, at the top $\bullet$ position in the table. This does indeed correctly produce $[1/2]$ at the third $\bullet$ position. The $[N/N + 1]$ approximants are useful in various quantum mechanical problems involving resolvent operators and sum rules. Wynn's algorithm gives directly the *numerical value* of the approximants and does not produce the numerator and denominator polynomials themselves. It is so simple that it will easily pack into a brief microcomputer program. Wynn and other workers have looked at other algorithms for the $[M/N]$, some of them with better stability properties, but for most problems Wynn's original method works satisfactorily. I find that the stability of the algorithm is criticised by some authors and praised by others. For example, it looks as though a small $(C - B)$ value will cause trouble, but in many cases a small $(C - B)$ will mean that we have reached convergence, so we will be stopping the calculation anyway! In any case a large value in the odd numbered columns comes out as a small correction term in the approximant columns (see solution 1 for a further comment). Wynn [6] has looked at ways of avoiding division by zero in various algorithms similar to his own.

Exercises

1. Write a program to implement Wynn's algorithm and work out some higher order approximants for the Euler series at $z = 0.1$, taking the sums of the series from the table in the text.
2. Using the Schwartz inequality

$$|\langle f|g \rangle|^2 \leqslant \langle f|f \rangle \langle g|g \rangle \tag{25}$$

    find a simple necessary condition for a series to be a series of Stieltjes.
3. In §3.4 I pointed out the iterative prescription

$$y = 2x - Mx^2 \tag{26}$$

    for finding M^{-1}. Setting $M = (1 + zt)$, and working out $\langle \psi|y|\psi \rangle$ as our estimate of $F(z) = \langle \psi|(1 + zt)^{-1}|\psi \rangle$, we get

$$F(z) = 2\langle \psi|\phi \rangle - \langle \phi|(1 + zt)|\phi \rangle \tag{27}$$

    where $\phi = x\psi$ is some arbitrary trial function. Show that the quantity on the right is a *lower bound* to $F(z)$ if z is real and positive. Try the trial functions $\phi = k\psi$ (with variable k), $\phi = k(1 - zt)\psi$ (with variable k) and $\phi = k(1 - \beta zt)\psi$ with variable k and β. Optimise with respect to k and see what estimate of $F(z)$ results.
4. Obtain a power series solution to the differential equation

$$x^2 Dy = x(x - 1) + y. \tag{28}$$

    Show that $y = e^{-x^{-1}} + x$ obeys the equation and compare it with your solution.
5. Write the quantity $(1 + \lambda)^{1/2}$ as $1 + h$ and obtain an infinite continued fraction for h. Work out the fraction to various levels and see how the results compare with $\sqrt{1.2}$ at $\lambda = 0.2$.
6. The exponential series is not a Stieltjes series (solution 2). Work out some Padé approximants for it to see whether they appear empirically to converge to give a good e^z value for small z.
7. The input-output formula

$$y = \tfrac{1}{2}[x + Mx^{-1}] \tag{29}$$

    gives a second-order iterative process for the square root $M^{1/2}$. Setting

$M = (1 + X)$, with starting input $x_0 = 1$, can you see what sequence of Padé approximants result for $(1 + X)^{1/2}$?

Solutions

1. One possible program is as follows (in Pet style), for $\Sigma A(n)z^n$

```
 5 INPUT T
10 DIM E(T, T), P(T, T), A(T)
15 FOR M = 0 TO T : E(M, 0) = 0
20 PRINT "A", M : INPUT A(M) : NEXT M
25 PRINT "Z" : INPUT Z
30 Q = T : S = A(0) : N = 0 : L = 1 : P(0, 0) = A(0)
35 FOR M = 0 TO Q : L = Z * L
40 S = S + A(M) * L
45 P(M, 0) = S : NEXT M
55 P(0, 1) = A(0) * A(0)/(A(0) − Z * A(1))
60 N = N + 1 : Q = Q − 1 : IF N > Q + 1 THEN 25
65 FOR M = N − 1 TO Q
70 E(M, N) = E(M, N−1) + 1/(P(M + 1, N − 1) − P(M, N − 1))
75 NEXT M : NN = N − 1 : IF NN = 0 THEN NN = 1
80 FOR M = NN TO Q
85 P(M, N) = P(M, N − 1) + 1/(E(M, N) − E(M − 1, N))
86 IF M = N OR M = N + 1  THEN 90
88 GOTO 95
90 PRINT P(M, M)
95 NEXT M : GOTO 60
```

The PRINT statements can be varied; any required set of the P(M, N) i.e. [M/N] can be printed in line 90. The GOTO 60 asks for another z value without us having to put all the coefficients in again. T is the maximum power, e.g. 8 if we go up to the z^8 coefficient. The E(M, N) are the numbers in the odd rows. Line 55 puts in specially the value of [0, 1], as I explain in the text. The various tricks with Q and NN are just to ensure that the columns of numbers contract in length properly as we move across the table.

The results for the Euler series case are (at $z = 0.1$) $[9/2] = 0.915\,633\,107$, $[8/2] = 0.915\,633\,636$, which give

$$F(z) = 0.915\,633\,37 \pm 2.7 \times 10^{-7}.$$

From [8/3] and [7/3] we find

$$F(z) = 0.915\,633\,347 \pm 3.5 \times 10^{-8}$$

and from [6/5] and [5/5]

$$F(z) = 0.915\,633\,341 \pm 4 \times 10^{-9}$$

Thus the error has been reduced by a factor of 50 000 or so by using the $[M/N]$ with $N \neq 0$ but still using only the same number of terms. In fact we are here at a stage where the results are so good that we have to worry about rounding error in the last digits! We note, as an amusing example that the integral

$$\int_0^\infty \exp(-2r)r^2(1+zr)^{-1}\,dr \tag{30}$$

which represents $\langle (1+zr)^{-1} \rangle$ for the hydrogen atom ground state, could be estimated accurately from a Padé analysis using only the well known $\langle r^n \rangle$ for positive n. The series would be the divergent one

$$\langle (1+zr)^{-1} \rangle = \tfrac{1}{8} \sum_0^\infty \left(\frac{-z}{2}\right)^n (n+2)! \tag{31}$$

2. With $f = t^N \psi$ and $g = t^M \psi$ we get

$$\mu_{M+N}^2 \leqslant \mu_{2M}\mu_{2N}. \tag{32}$$

For $N = (K+1)/2$ and $M = (K-1)/2$ this becomes

$$\mu_K^2 \leqslant \mu_{K-1}\mu_{K+1} \tag{33}$$

as one simple necessary condition. The exponential series fails the test!

3. If we use any function of form $\phi = kf\psi$, with k variable, we get a result of form $2ak - bk^2$ for $F(z)$. This has an extremum with a value a^2/b, so we can eliminate k at once. With $f = 1$ we get the $[0/1]$ approximant for $F(z)$. With $f = (1-zt)$ we get a rational $[0/1]$ approximant for a regrouped version of the series, as though the first two terms were

$$F(z) = (\mu_0 - z\mu_1) + (\mu_2 z^2 - \mu_3 z^3) = T_0 + T_1. \tag{34}$$

The approximant is $T_0^2/(T_0 - T_1)$. For $f = (1-\beta zt)$ we get the approximant N/D with

$$N = (\mu_0 - \beta z\mu_1)^2 \tag{35}$$

$$D = (\mu_0 + z\mu_1) - 2\beta(z\mu_1 + z^2\mu_2) + \beta^2(z^2\mu_2 + z^3\mu_3). \tag{36}$$

By optimising with respect to β as well as k we get an approximant which we can describe as 'the minimum value of $N(\beta)/D(\beta)$ as β varies'. It is the $[1/2]$ Padé approximant; even though N/D *looks* like the ratio of polynomials of degrees 2 and 3 the optimum β is a series in z such that collecting the powers of z together gives the [1/2] approximant. Indeed, by using the trial function $\Sigma_0^N A(n)(zt)^n$ in the variational principle and optimising all the A_n we get the $[N/N+1]$ Padé approximant, although to prove this algebraically is a tedious job.

4.   Setting $y = A_0 + A_1 x + \ldots$ into the equation yields the result $y = x$, with all other A_n zero. The function $e^{-x^{-1}}$ is singular at $z = 0$ and an attempt to obtain a power series for it by regarding x as real and letting $x \to 0$ yields a series $0 + 0x + 0x^2 + \ldots$. Thus any power series representing a function about $x = 0$ will tell us nothing about possible 'invisible' component functions of this type. The usual way of saying this is that a function has only one asymptotic series for real positive z, but one asymptotic series can represent many functions. An asymptotic series for a function $f(z)$ has (in Poincaré's definition) the kind of common-sense property that we expect of a convergent series: if we take the sum S_N to the z^N term we will have a z^{N+1} term next, so the quantity $|f(z) - S_N|/|z|^N$ will tend to zero with $|z|$. This property is what defines an asymptotic series for $f(z)$, but it can be obeyed by a divergent series (as for our Euler example) or by a null (zero) series (as for $e^{-z^{-1}}$). If we know that the series represents a function, such as a function of Stieltjes type, which is analytic at $z = 0$, then we can ignore the possibility of hidden components. Perturbation theory in quantum mechanics uses power series, of course, and it often gives us only an asymptotic series [7, 8]. This can even happen when the series is a convergent one with only a finite number of non-zero terms [§7.3]. However, the use of Padé and other methods has tamed some unruly series in recent years, and so there is still room in the subject for both the pessimists and the optimists!

5.   Setting $(1 + \lambda)^{1/2} = 1 + h$ gives the result

$$h = \lambda/(2 + h) \tag{36}$$

$$= \cfrac{\lambda}{2 + \cfrac{\lambda}{2 + \ldots}}$$

$$= \frac{\lambda}{2} + \frac{\lambda}{2} + \frac{\lambda}{2} + \ldots \tag{37}$$

if we use a 'linear' way of writing the infinite continued fraction. Consider now the more general continued fraction

$$F = \frac{a_1}{b_1} + \frac{a_2}{b_2} + \frac{a_3}{b_3} + \ldots \tag{38}$$

By ignoring every a_n beyond a given one we get an estimate of F called a *convergent* to the continued fraction. Thus the first convergent to F is a_1/b_1 and the second is

$$\frac{a_1}{\left[b_1 + \dfrac{a_2}{b_2}\right]} = \frac{b_2 a_1}{b_2 b_1 + a_2} \, . \tag{39}$$

In general each convergent is a fraction with numerator A_k and denominator B_k. There is a simple recursion relation between the A_ks:

$$A_k = b_k A_{k-1} + a_k A_{k-2}. \tag{40}$$

The *same* recursion relation holds for the B_k, if we use the initial values $A_{-1} = 1$, $B_{-1} = 0$, $A_0 = 0$, $B_0 = 1$. The proof of the result is by induction and the most elegant version which I know is that in Hall and Knight's *Higher Algebra,* first published in 1887. Just as a polynomial may be worked out forwards or backwards (exercise 2.6) so may a continued fraction, and Jones and Thron [9] have discussed the two approaches. For our example, with $\lambda = 1$, we can apply the recursion relations to get a sequence of convergents. Alternatively, we can keep λ and get a sequence of rational fractions in λ which turn out (as the suspicious reader might have been expecting) to be Padé approximants: they are actually [1/0], [1/1], [2/1], [2/2], etc. The third and fourth convergents give $\sqrt{1.2} = 1.095\,450 \pm 5 \times 10^{-6}$. That the successive convergents straddle the exact result follows from the classical theory of continued fractions [10]. There is a close link between that theory and the theory of functions of Stieltjes type. The main point is that a study of the series of Stieltjes can be converted into the study of a continued fraction of type

$$\frac{a_1}{1} + \frac{a_2}{1} + \dots$$

with the $a_k > 0$, for which the various straddling properties are already part of traditional mathematics. Some authors proceed by getting a series, converting it to a continued fraction by some algorithm, and then working out the convergents by a recursive algorithm such as that which I described above. Clearly, if we only want numbers it is quicker to do the job at one go by using an approach such as Wynn's algorithm. However, those readers interested in continued fractions will find in a paper by Gordon [11] a concise recursive method which constructs the fraction from the given series and which is almost as simple as Wynn's algorithm. Having got the fraction we can get the convergents using the recursion relation quoted above. An interesting review of the history of the theory, particularly as relating to the continued fractions, is given by Shohat and Tamarkin [12].

6. The first problem you will encounter is that the computer keeps stopping because of divisions by zero. This arises for small z because [7/0] and [8/0] might be equal, say, owing to quick convergence of the series. Several authors have written papers about how to modify the Wynn algorithm to allow for zero divisors. I approached the problem by arguing that a divisor of 10^{-10}, say, would presumably not be much different from 0 as far as the rest of the calculation is concerned, but it wouldn't stop program execution. Suppose that we want to calculate $B = A + 1/(P1 - P2)$ and would get a division by zero. I tried using the steps

$$D = (P1 - P2)$$
$$\text{IF } D = 0 \text{ THEN } D = 1E - 10$$
$$B = A + 1/D$$

In the cases which I have tried this trick allows the calculation to continue without spoiling later values in the table (i.e. when all columns converge they go to the same limit, which is the correct e^z value for small z). For $z = -4$ the $[N/N]$ approximants converge more quickly than the $[2N/0]$ ones and give a result for e^z in error by -6×10^{-10} on the Pet. However, the straddling property of alternate approximants is not always present, whereas it *has* to be for a series of Stieltjes. I note in passing the $[2/2]$ approximant, 'all the threes', which is useful for simple calculator work [13]:

$$e^z \cong \frac{(3 + z)^2 + 3}{(3 - z)^2 + 3} \tag{41}$$

7. Input 1 gives output $1 + \frac{1}{2}X$, the $[1/0]$ approximant, with an error of order X^2. Repeating the process will give an error of order X^4 i.e. it will fit the series for $(1 + X)^{1/2}$ up to the X^3 term. Since the denominator is $(1 + \frac{1}{2}X)$, the result will be the $[2/1]$ approximant, and it equals $[1 + X + \frac{1}{8}X^2]/[1 + \frac{1}{2}X]$. At the next stage the error is of order X^8 and the $[4/3]$ approximant results; the next stage gives the $[8/7]$ approximant; the general sequence is thus formed by the $[2^N/2^N - 1]$ approximants. For any positive X the sequence of approximants converges to $(1 + X)^{1/2}$, although the power series diverges for $X > 1$.

Notes

1. E T Bell 1953 *Men of Mathematics* (Harmondsworth: Penguin)
2. A Erdelyi 1956 *Asymptotic Series* (New York: Dover)

3. B W Roos 1969 *Analytic Functions and Distributions in Physics and Engineering* (New York: John Wiley)
4. S Graffi and V Grecchi 1978 *J. Math. Phys.* **19** 1002
5. P Wynn 1956 *Math. Tables and Aids to Computation* **10** 91
6. P Wynn 1963 *B.I.T.* **3** 175
7. J B Krieger 1968 *J. Math. Phys.* **9** 432
8. R G Wilson and C S Sharma 1980 *J. Phys. B: At. Mol. Phys.* **13** 3285
9. W B Jones and W J Thron 1970 *Math. Comput.* **28** 795
10. A Ya Khinchin 1964 *Continued Fractions* (Chicago: University of Chicago Press)
11. R G Gordon 1968 *J. Math. Phys.* **9**, 655
12. J A Shohat and J D Tamarkin 1963 *The Problem of Moments* (Am. Math. Soc.)
13. J P Killingbeck 1981 *The Creative Use of Calculators* (Harmondsworth: Penguin)

7 A simple power series method

7.1 Introduction

Chapter 6 was concerned with the treatment of power series which are divergent, with the aim of getting a useful 'sum' for the series. In chapter 9 I describe a way of calculating divergent perturbation series for some quantum mechanical problems and also introduce another way to produce a sum for them. In this chapter, by contrast, I look at a class of problem for which the series concerned are convergent, in the sense of formal mathematics, but give slow convergence when they are simply 'added up' directly. By using a combination of two simple mathematical tricks I show how to speed up markedly the rate of convergence of the method when it uses the wavefunction series to estimate the eigenvalue. The resulting method of eigenvalue calculation is one of the most simple and accurate ones available for a microcomputer, and I apply it to the charmonium problem and to the quadratic Zeeman problem in chapter 12. As a necessary preliminary to explaining the method I set out some useful forms of the one-particle Schrödinger equation, including a slightly unusual form of the radial equation which turns out to be particularly appropriate for use with the finite-difference methods of chapter 10. As I pointed out in chapter 1, it is always useful to have available a few test problems, with known exact solutions, in order to test any proposed numerical method. §7.3 gives a few examples of test problems which can be used to test methods of eigenvalue calculations; it also gives a perturbation problem for which the energy series, although convergent, does not correctly give the perturbed energy.

7.2 Standard forms of the Schrödinger equation

As I explained in §2.5 it is usual in computational work to drop quantities such as h, m, and e from the Schrödinger equation and to treat it in some simple reduced form. In the scientific papers dealing with the one-dimensional Schrödinger equation, particularly that for perturbed oscillator systems, the kinetic energy operator often takes the form $-D^2$. The harmonic oscillator Schrödinger equation is then

$$-D^2\psi + x^2\psi = H\psi = E\psi \tag{1}$$

and by direct trial we can see that $\psi_0 = \exp(-\tfrac{1}{2}x^2)$ obeys this equation with $E = 1$. To get higher energy states we can act repeatedly on ψ_0 with the shift operator $\eta = x - D$, which obeys the commutator relation

$$[H, \eta] = 2\eta \tag{2}$$

and so increases the eigenvalue by 2 each time it acts. Some standard textbooks use shift operator algebra for the oscillator problem. The approach can be applied for potentials other than x^2; it is then usually called the Infeld–Hull factorisation technique and the shift operator η becomes very complicated (for a recent example see [1]). For the oscillator problem it is clear that each excited state wavefunction is equal to ψ_0 multiplied by a polynomial, and many textbooks adopt a power series approach to the solution of the oscillator problem. (The polynomials are, of course, the well known Hermite polynomials.) As I shall show in §7.4 it is the power series method which is the most powerful one for microcomputer work, although the shift operator technique has a strong appeal to physicists who like operator methods. I urge such readers to look at the delightful and unusual quantum mechanics text by Green [2]).

The boundary conditions $\psi = 0$ at $x = \pm\infty$ are often used for one-dimensional bound state problems, but for *radial* problems the boundary conditions usually involve assigning ψ at $r = 0$ and $r = \infty$. In much of the current research literature the hydrogen atom Schrödinger equation is taken in the reduced form

$$-\tfrac{1}{2}\nabla^2\psi - r^{-1}\psi = E\psi \tag{3}$$

with the ground state wavefunction $\exp(-r)$ having $E = -\tfrac{1}{2}$. To get wavefunctions of angular momentum l the traditional route is to set $\psi = Y_l^m(\theta, \phi)r^{-1}R(r)$, where Y is a spherical harmonic. The resulting equation for the function $R(r)$ is then usually called the radial equation and takes the form

$$-\tfrac{1}{2}D^2R + V(r)R + \tfrac{1}{2}l(l+1)r^{-2}R = ER \tag{4}$$

where $-r^{-1}$ is replaced by an arbitrary central potential $V(r)$. The equation looks like a one-dimensional one but has an extra centrifugal potential term. The boundary conditions for bound states are $R = 0$ at $r = 0$ and $r = \infty$. There is another way to approach the problem, however, which I have found [3] to be useful for microcomputer work, and which as far as I know has only been used by a few workers on angular momentum theory. The idea is to set

$\psi = Y_l\phi(r)$, where Y_l is a *solid* harmonic of degree l (see Appendix 1), so that the angular momentum is l. After a little algebra using the identity

$$\nabla^2 (fg) = f\nabla^2 g + g\nabla^2 f + 2 \operatorname{grad} f \cdot \operatorname{grad} g \tag{5}$$

we arrive at an equation for ϕ;

$$-\tfrac{1}{2}D^2\phi - (l+1)r^{-1}D\phi + V\phi = E\phi. \tag{6}$$

If we further set $R = r^{l+1}\phi$ we get back to the traditional radial equation for R. However, the ϕ equation takes a very simple form in finite-difference language (§10.6) and does not have a centrifugal potential term.

7.3   Some interesting test cases

The usual approach to the Schrödinger equation is to give the potential function V and then try to find the energies and the eigenfunctions. However, to obtain simple test problems (with which to test various computational techniques) it is easier to start with a wavefunction and derive the energy and the potential. For example, using the wavefunction $\psi = \exp\,[-f(x)]$ gives the result

$$-D^2\psi = [f'' - f'^2]\psi. \tag{7}$$

The choice $f = \tfrac{1}{2}x^2 + \lambda x^4$ then gives

$$-D^2\psi + [x^2(1 - 12\lambda) + 8\lambda x^4 + 16\lambda^2 x^6]\psi = 1\psi. \tag{8}$$

To keep ψ normalisable, λ (or at least its real part) must be positive, and the eigenvalue 1 is then independent of the magnitude of λ. For $\lambda < 0$ the function ψ is not normalisable, but the $\lambda^2 x^6$ term ensures that the Schrödinger equation will still have bound states! For small negative λ there will be an eigenvalue very close to 1, but it moves away from 1 as $|\lambda|$ increases. For large positive λ the potential is a deep double well one, with a local maximum at $x = 0$; such double well potentials are favourite ones in the literature for providing difficult test cases to compare the merits of different techniques of eigenvalue calculation [4].

   As an example for the three-dimensional equation I note the choice $\psi = \exp\,(-r - \lambda r^2)$. This obeys the equation

$$-\tfrac{1}{2}\nabla^2\psi + (2\lambda r + 2\lambda^2 r^2 - r^{-1})\psi = (3\lambda - \tfrac{1}{2})\psi. \tag{9}$$

For $\lambda < 0$ the function ψ is not normalisable but the Schrödinger equation still has a bound state with energy close to $(3\lambda - \tfrac{1}{2})$ when $|\lambda|$ is small [5]. The Rayleigh-Schrödinger perturbation series for the energy, based on the unperturbed Coulomb potential $-r^{-1}$, is just $-\tfrac{1}{2} + 3\lambda$, with the higher order coefficients vanishing identically. For $\lambda > 0$ this finite series gives the eigenvalue exactly, but for $\lambda < 0$ it is in error by an amount which increases rapidly with $|\lambda|$.

7.4   The power series approach

For both the harmonic oscillator and the hydrogen atom the eigenfunctions take the form of an exponential factor multiplied by a polynomial. However, it is not the finite number of terms which matters; for example, the ground state function $R(r) = r \exp(-r)$ for the hydrogen atom radial equation is really an infinite power series if we expand out the exponential. The point is that the series converges *and* gives a function which is normalisable, in the sense that R^2 integrated over all space gives a finite result. The function $r \exp(r)$, for example, also has a convergent power series at every r but does *not* give a normalisable function. I will illustrate the idea of the power series approach by looking at a celebrated problem which has been treated in scores of research papers, namely the perturbed oscillator problem

$$-D^2 \psi + (\mu x^2 + \lambda x^4)\psi = E\psi. \tag{10}$$

The potential function is a convergent (indeed finite) series around $x = 0$. For a given E the basic theorems about differential equations (due to Frobenius) tell us that we can thus find a convergent power series in x for the wavefunction $\psi(x, E)$. In principle, then, at some large x value we can get $\psi(x, E)$ by adding up a sufficient number of terms of the series. To get a bound state, though, we want $\psi(x, E)$ to tend to zero as x tends to infinity; this second requirement is *additional* to the convergence requirement and it can only be satisfied for specific E values, the required eigenvalues. The calculational tactics to use are clear. We pick on some large x and take two trial energies E_1 and E_2. Using the series for $\psi(x, E)$ we take sufficient terms to get converged values for $\psi(x, E_1)$ and $\psi(x, E_2)$. Then by linear interpolation we estimate the E value which would have made $\psi(x, E)$ zero. We then repeat, using $E_1 = E - D$, $E_2 = E + D$, with D small. After a few repetitions we should get a close estimate of the eigenvalue. That is the idea; what spoils it in practice? Well, it was tried a few years ago [6] and it worked quite well on some problems. However, it sometimes turned out that thousands of terms of the series had to be taken before a converged result was obtained. Not only does this take a long time; it also allows rounding errors to accumulate so that the final ψ estimates and thus the eigenvalue estimate are rendered unreliable. I discovered recently that a sure empirical sign that is happening is the appearance of fluctuations in the E value obtained on successive runs with slightly varied starting estimates E_1 and E_2. I managed to cure the problem in most cases by using a convergence factor in the wavefunction and by studying a wavefunction *ratio*; the number of terms of the series needed was then reduced by a factor of up to twenty [7]. With this recent improvement the power series approach becomes one of the simplest and most accurate methods for a microcomputer. I return to the perturbed oscillator problem to explain the details. The trick (a very simple one) is to write the wavefunction

ψ in the form

$$\psi = \exp(-\beta x^2)F. \tag{11}$$

Putting this into the Schrödinger equation produces the equation

$$-D^2F + 4\beta xDF + (\mu x^2 - 4\beta^2 x^2 + \lambda x^4)F = (E - 2\beta)F. \tag{12}$$

This equation for F could be treated by several techniques, but the power series approach sets

$$F = \sum_0^\infty A(n)x^n = \sum_0^\infty T(n). \tag{13}$$

By putting this form of F into the equation we get a recurrence relation

$$(N+1)(N+2)T(N+2) = (4\beta N + 2\beta - E)T(N)x^2$$
$$+ (\mu - 4\beta^2)T(N-2)x^4 \tag{14}$$
$$+ \lambda T(N-4)x^6.$$

We can either take $T(0) = 1$, with all the Ns even, to get an even solution, or take $T(1) = 1$, with all the Ns odd, to get an odd solution. Since only four coefficients appear at a time in the calculation we don't even need to officially call the $T(N)$ an array, but could call them A, B, C, D, for example. The following simple BASIC program (in Pet style) will do the calculation in the manner outlined in the preceding discussion.

```
10 INPUT N, X, L, M
20 INPUT E, DE, P
30 F1 = 1 : F2 = 1 : C1 = 0 : C2 = 0 : D1 = 0 : D2 = 0
40 B1 = 1 : B2 = 1 : X2 = X ↑ 2 : X4 = X ↑ 4 : X6 = X ↑ 6
45 L = L * X6 : K = (M − 4 * P * P) * X4 : Q = 1
50 J1 = (2 * P − E) * X2 : J2 = J1 − DE * X2
60 D = (N + 1) * (N + 2) : T = 4 * N * P * X2
70 A1 = (T + J1) * B1 + K * C1 + L * D1
80 A2 = (T + J2) * B2 + K * C2 + L * D2
90 A1 = A1/D : A2 = A2/D : Q = 1
95 1F ABS (F1) > 1E 30 THEN Q = 1E − 6
100 F1 = (F1 + A1) * Q : F2 = (F2 + A2) * Q
110 EP = E + DE/(1 − F2/F1)
120 D1 = C1 * Q : C1 = B1 * Q : B1 = A1 * Q
130 D2 = C2 * Q : C2 = B2 * Q : B2 = A2 * Q
140 PRINT EP : N = N + 2 : GOTO 60     (fixed print position)
```

Of course, it is equally possible to make the quantities into arrays and use lots of FOR-NEXT loops to do the manipulations. The quantities in the E1 calculation are called A1, B1, etc, while those in the E2 calculation are A2, B2, etc. E2 is represented as E1 + DE. By setting N = 0 or 1 we pick out even or odd states, respectively. The projected energy EP is worked out in line 110. By using the T_n instead of the A_n we reduce the possibility that the factor x^n in T_n will cause overflow even though T_n is not large enough to do so. Overflow can be controlled in most cases by the statements in lines 90 and 95.

As an example I take the case $\mu = 0, \lambda = 1$, for which we have an approximate eigenvalue 50.2 from the WKB approximation (§5.7). This particular example involves summing many terms if F1 is worked out with $\beta = 0$ [6]; however for any β between 3 and 10 the number of terms needed is reduced drastically and an accurate eigenvalue is obtained. Although it takes a little while to find an appropriate β value, there is usually quite a wide range of β over which good results can be obtained, and the operator's skill at estimating a likely value improves with experience. The table below sets out the results of one sequence of runs, with X = 6. Using X = 7 gives no change, so we can take it that X = 6 is large enough to be effectively giving us the boundary condition $\psi(\infty) = 0$. N is the rough N value in the series needed to give convergence: β is held at the value 5.

E_1	DE	N	EP
50.0	0.2	250	50.25
50.25	0.02	200	50.2562
50.2562	0.0002	140	50.256 2545
	0.0001	140	50.256 2545

Using $\beta = 0$ at the last stage requires an N of over 400 and also produces a result which differs for the two different DE values, illustrating the value of using a non-zero β. At the optimum β (about 3) only an N value of 80 is needed at the last stage. One disadvantage of the series approach is that we do not know *which* eigenvalue we have obtained if we have no further information. In this case we know that it is the $n = 10$ state, since we started from a WKB estimate. However, we are able to get excited state energies without first treating lower states, whereas in a matrix-variational approach we have to set up a basis of at least eleven functions to get at the $n = 10$ state. The interested reader who uses printout lines for F1 and F2 will discover that they only stabilise at around N = 800 for the last stage of the above calculation! That is why I chose the ratio F2/F1 as the thing to use to get the projected energy PE; the ratio converges long before the individual wavefunctions do. In [6] F1 and F2 were calculated one after the other, so that N values of thousands were needed, since $\beta = 0$ was used. Clearly a little analysis has served to improve speed and accuracy quite markedly for this problem! Any polynomial potential in powers of x^2 can be treated by simply including the appropriate terms in the recurrence relation for the T_N.

Readers with sufficient experience to have developed their own programming style may not agree with every detail in my program, but I think that any effective program for the power series method must conform fairly closely to the following flowchart.

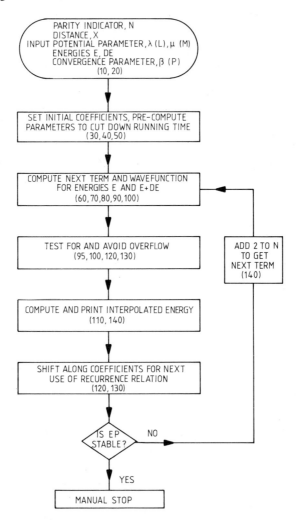

Exercises

1. Look for solutions of form exp $(-kr)$ for the H atom ϕ function of equation (6). Find energy levels for the unperturbed harmonic oscillator by setting

$\lambda = 0$ in the recurrence relation (14) and then choosing β appropriately, with $\mu = 1$.

2. Study the oscillator problem with $\mu = -10$, $\lambda = 1$. Use the series method program to look for even and odd states with energies near to -20 and -12.

3. The series method can be used to find energy levels obeying the homogeneous Dirichlet boundary condition $\psi(\pm X) = 0$ for any X value; for large X the energy values tend to a limit which corresponds to the common bound state conditions $\psi(\pm \infty) = 0$. Homogeneous Neumann boundary conditions take the form $D\psi(\pm X) = 0$, i.e. they involve the gradient of ψ. How can the series method be used to handle such boundary conditions?

4. Suppose that some given analytical function ψ is an approximate ground state function for the Hamiltonian $H = -D^2 + V$, and that the expectation value $\langle \psi | H | \psi \rangle$ is ϵ. How can we produce a potential U such that ψ is an exact eigenfunction for the Hamiltonian $-D^2 + U$, with exact eigenvalue ϵ? Suppose that ψ takes the (unnormalised) form $\exp[-\beta x^2 - \gamma x^4]$. How can we choose β and γ so as to make ψ an approximate eigenfunction for the Hamiltonian $-D^2 + x^2 + \lambda x^4$ when λ is a small positive number?

5. Consider the differential equation

$$D^2 y = x^\beta y \tag{15}$$

studied by Bender and Sharp [8] for the cases $\beta = 1$ and $\beta = 2$. $y(0)$ can be taken to be 1, but the problem is to choose the value of the slope $Dy(0)$ so that the solution function tends to zero as x tends to infinity (using the region $0 \leqslant x < \infty$). Devise a series solution method for finding $Dy(0)$ and apply it to the case $\beta = 1$.

Solutions

1. Setting $\exp(-kr)$ into the equation for ϕ shows that the choice $k = (l+1)^{-1}$ gives an eigenfunction with $E = -\frac{1}{2}(l+1)^{-2}$ and, of course, the total wavefunction $Y_l \phi$ has angular momentum l. The states concerned are those traditionally labelled 1s, 2p, 3d, etc, being the lowest energy states for their particular l value.

    With $\lambda = 0$, $\beta = \frac{1}{2}$, the recurrence relation for the T_n shows that if $E = 2N + 1$ for integer N then the coefficient A_{N+2} is zero. The wavefunction is then normalisable and is a polynomial multiplied by the function $\exp(-\frac{1}{2}x^2)$.

2. The potential is a deep double well and so functions (L and R) centred on the left and right well are almost decoupled from one another. To get even

and odd states we must form the combinations $L \pm R$ and the energy splitting between the even and odd states depends on the matrix element $\langle L|H|R \rangle$ where H is the Hamiltonian (energy operator). The above discussion in terms of the matrix-variational approach is not directly relevant to the series method, which just ploughs ahead and gives the energy values without using matrix methods directly (although it is related to them, as I shall point out in §8.3). The (even, odd) level pairs are $(-20.633\,5767, -20.633\,5468)$ and $(-12.379\,5437, -12.375\,6738)$.

3.   The total wavefunction is a product $\psi = \exp(-\beta x^2)F$. To make ψ zero it is sufficient to make F zero, which is what the series method aims at. However to make $D\psi = 0$ we have to use the result

$$D\psi = \exp(-\beta x^2)[F' - 2\beta xF] \tag{16}$$

and so have to make the quantity in square brackets zero. This simply involves calculating two sums,

$$F = \Sigma\, T(n) : F' = \Sigma\, nT(n) \tag{17}$$

and finding $G' = (F' - 2\beta xF)$. The quantities $G'(E_1)$ and $G'(E_2)$ are then used in place of $F(E_1)$ and $F(E_2)$ in the portion of the program which calculates the projected energy EP. Only a few extra statements need be added to the original program.

4.   We just work out $\epsilon\psi - D^2\psi$ and divide it by ψ to get U, which is often called the Sternheimer potential. If ψ is nodeless, as we choose it to be for a ground state problem with a local potential V, then the division by ψ causes no problem. Taking $(V - U)$ as a perturbation, it is clear that this perturbation has zero expectation value for ψ.

Putting $f = \beta x^2 + \gamma x^4$ into equation (7) gives the result

$$-D^2\psi + U\psi = 2\beta\psi \tag{18}$$

where

$$U = (4\beta^2 - 12\gamma)x^2 + 16\beta\gamma x^4 + 16\gamma^2 x^6 \tag{19}$$

To make this fit to the potential $x^2 + \lambda x^4$ we must have

$$1 = 4\beta^2 - 12\gamma : \lambda = 16\beta\gamma$$

These equations have a solution with $\beta \simeq \frac{1}{2}$ when λ is small, so that $16\gamma^2$ is roughly $\frac{1}{4}\lambda^2$. The quantity $(V - U)$ is thus about $-\frac{1}{4}\lambda^2 x^6$ when λ is small, and the ground state energy is a little lower than 2β when $(V - U)$ is allowed for.

5.   If we set

$$y = \exp(-kx) \Sigma\, A_n x^n \tag{20}$$

we have the initial conditions $A_0 = 1$, $A_1 = k + Dy(0)$. Using the notation $T_n = A_n x^n$ we obtain the recurrence relation

$$k^2 T_N R^2 - 2k(N+1)T_{N+1}R + (N+2)(N+1)T_{N+2} = T_{N-\beta}R^{\beta+2}$$

(21)

where R is the large distance at which we want the solution function to be zero. The procedure is similar to that for eigenvalues in §7.4, except that we use trial values of $Dy(0)$ instead of trial energies. For $\beta = 1$, using a value of $R = 20$, I obtained a value of $-0.729\,011\,132$ for $Dy(0)$. Varying R a little gives no change in the result, so $R = 20$ must be large enough to be effectively at infinity.

Notes

1. N Bessis, G Bessis and G Hadinger 1980 *J. Phys. A: Math. Gen.* **13** 1651
2. H S Green 1968 *Matrix Methods in Quantum Mechanics* (New York: Barnes and Noble)
3. J Killingbeck 1977 *J. Phys. A: Math. Gen.* **10** L99
4. B G Wicke and D O Harris 1976 *J. Chem. Phys.* **64** 5236
5. J Killingbeck 1978 *Phys. Lett.* **67**A 13
6. D Secrest, K Cashion and J O Hirschfelder 1962 *J. Chem. Phys.* **37** 830
7. J Killingbeck 1981 *Phys. Lett.* **84**A 95
8. C M Bender and D H Sharp 1981 *Phys. Rev.* **D24** 1691-94

8 Some matrix calculations

8.1 Introduction

Throughout this book I champion methods which use iterative calculations and recurrence relations. In chapter 3 some of the iterative methods which I described were for matrix calculations, while in chapter 7 I attacked a recurrence relation problem by using a power series method. In this chapter I return to the perturbed oscillator problem of chapter 7, but attack it by a matrix method which is iterative in spirit. Much of the modern literature of applied matrix theory deals with so-called sparse matrices, in which only a small fraction of the elements is non-zero. Such matrices often arise quite naturally in connection with problems in physics, and in §8.3 I give a detailed treatment of the perturbed oscillator problem as an example. The recurrence relation method which I describe is of fairly wide applicability and I use it again in chapter 10 in a discussion of one-dimensional band theory. In this book I concentrate mainly on matrix problems involving special types of sparse matrices, since these can be handled by microcomputers. On large computers, however, it is quite common for multi-stage computations to be performed; these involve numerical integration to find the matrix elements, followed by a matrix diagonalisation to get the eigenvalues. §8.2 gives a general discussion of the use of matrices, in an attempt to put the simple methods of this book in some perspective.

8.2 Matrices in quantum mechanics

The original version of matrix mechanics is little used in applications nowadays (but see [1, 2] for exceptions). In that early theory position, momentum and other quantities were represented by infinite matrices, whereas many modern applications of quantum mechanics represent such observables as operators (usually in a Hilbert space of normalisable wavefunctions). To capture in a matrix formalism such simple commutator properties as

$$[D, x] = Dx - xD = 1 \tag{1}$$

(i.e. the momentum-position commutator without its $i\hbar$) it is essential to use infinite matrices. For any two finite $N \times N$ matrices A and B we can establish by explicit computation that Trace (AB) = Trace (BA), where Trace denotes the sum of the diagonal elements of the matrix. Since Trace $(AB - BA) = 0$, it cannot equal N, which it would have to do if A and B were to represent D and x and obey the position-momentum commutation rule, with 1 interpreted as the unit $N \times N$ matrix.

Infinite matrices are unwieldy to use and have some disturbing properties, e.g. they do not obey the associative rule [3]. By far the most commonly appearing matrix in quantum mechanics is the matrix of the energy operator as set up in some finite basis of trial functions. The idea behind forming such a matrix is as follows. Suppose that we wish to solve the Schrödinger equation

$$H\psi = E\psi \tag{2}$$

by using as our postulated ψ a linear combination of N basis functions;

$$\psi = \sum_{1}^{N} A_n \phi_n. \tag{3}$$

By setting this form of ψ into the Schrödinger equation and taking the inner product of the resulting equation with each ϕ_n in turn we arrive at the set of equations

$$\sum \langle m|H|n \rangle A_n = E \sum \langle m|n \rangle A_n \tag{4}$$

where we use a Dirac notation for the matrix elements and inner products. Many introductory textbooks emphasise the use of orthonormal bases for which $\langle m|n \rangle = \delta_{nm}$ and for which the right-hand side of the above equation becomes EA_m. This restriction is not essential; indeed in quantum chemistry is is mainly because basis orbitals on different atoms have non-zero inner products that chemical bonds are formed. If we use the matrix notation H and S for the energy and overlap matrices and A for the column of coefficients, then the system of equations becomes what is called a *generalised eigenvalue problem* in matrix theory:

$$HA = ESA. \tag{5}$$

(The ordinary eigenvalue problem has EA on the right-hand side). The N eigenvalues of this matrix problem give upper bounds to the energies of the lowest N bound states of the energy operator H, under the assumption that there are such states, of course. (For a careful discussion of this upper bound property see [4]; some subtleties which arise when H refers to an atom with few electrons are treated in [5].)

The condition that E shall be an eigenvalue of the generalised eigenvalue problem is that the determinant $D_N(E)$ of the $N \times N$ square matrix $(H - ES)$ shall be zero. Thus E can be found either by various matrix diagonalisation techniques or by a 'search and find' operation in which the determinant is worked out for various trial E values and interpolation is used to find the eigenvalues. In the second type of approach a procedure similar to that of Newton's method for polynomial equations can be used (§3.3) with E playing the role of x and $D_N(E)$ playing the role of $f(x)$. When the matrix methods outlined above are used the computational tasks involved can be put in order as follows:

(1) Choose a basis set.
(2) Work out the matrix elements of H and S. If analytical formulae are not available this task might involve explicit numerical integration.
(3) Use some technique to calculate the eigenvalues (and eigencolumns) of the resulting matrix problem.
(4) Increase the number of basis states used and repeat the calculation until the lower eigenvalues stabilise at limiting values which are taken to be the eigenvalues of the original Schrödinger equation. If the basis set is not complete it is possible that a pseudo-limit higher than the actual eigenvalue is obtained. For example, in calculations of the quadratic Zeeman effect (§12.4) the use of any number of hydrogenic bound state basis functions cannot succeed, since continuum components are also needed to give a complete basis set.

Clearly a calculation might involve many integrations and matrix manipulations, so a large computer with double precision arithmetic is needed for some applications. Much effort is being devoted in the computer journals nowadays to methods of diagonalising and inverting matrices by using only a small portion of the matrix at a time, so that really enormous matrices (with $N \cong 10\,000$) can be handled by feeding a few elements at a time into the computer's fast store.

From the viewpoint of linear space theory what the Rayleigh-Ritz method does is to choose E so that the N inner products $\langle \phi_n | (H - E) | \psi \rangle$ are zero when ψ is a linear combination of the ϕ_n. What is *really* intended is to make $(H - E)\psi$ exactly zero. If it were zero then its inner product with *any* function would vanish. We could use N functions f_n which differ from the ϕ_n and set the N inner products $\langle f_n | (H - E) | \psi \rangle$ equal to zero; this gives the *Galerkin method*, which produces an $N \times N$ generalised matrix eigenvalue problem. The matrix elements $\langle f_m | (H - E) | \phi_n \rangle$ may be easy to calculate if the set f_n is chosen carefully, giving a computational advantage. As N is increased the matrix eigenvalues will tend to limits which (we hope) are the energy eigenvalues; however, for a given N the eigenvalues are not necessarily upper bounds to the true ones. (The Rayleigh-Ritz approach is equivalent to a variational approach and gives upper bounds.) If the f_n are taken to be localised at particular points in space then we get a so-called collocation method, which involves fitting the Schrödinger equation at a discrete set of points.

8.3   The Hill determinant approach

I now describe a matrix technique which does not involve taking inner products. If it turns out that the energy operator H and the basis ϕ_n have the property that $H\phi_m$ is a finite linear combination of the ϕ_n for any m, then we get a matrix problem directly without using inner products. I use the energy operator

$$H = -D^2 + \mu x^2 + \lambda x^4 \tag{6}$$

as an example, with the basis set $\phi_n = x^n \exp(-\beta x^2)$. We already know from §7.4 that $H\phi_m$ is a linear combination of only four different ϕ_n. If we set the eigenfunction ψ equal to a sum of the ϕ_n,

$$\psi = \Sigma A_n x^n \exp(-\beta x^2) \tag{7}$$

then we arrive at a recurrence relation linking four A_n at a time, just as in §7.4. The determinantal condition for finding the eigenvalues of even states (with n even) then takes the form that the determinant

$$\begin{vmatrix} d_0 & \alpha_0 & 0 & 0 & . \\ \beta_2 & d_2 & \alpha_2 & 0 & . \\ \gamma_4 & \beta_4 & d_4 & \alpha_4 & . \\ 0 & \gamma_6 & \beta_6 & d_6 & . \\ . & . & . & . & . \end{vmatrix} \tag{8}$$

must be zero, with

$$\alpha_N = -(N+1)(N+2)$$
$$d_N = (4\beta N + 2\beta - E) \tag{9}$$
$$\beta_N = (\mu - 4\beta^2) : \gamma_N = \lambda.$$

(For odd states we use α_{N+1} and d_{N+1} in place of α_N and d_N.)

I denote by D_N the determinant of the 'down to d_N' portion of this infinite determinant (often called a Hill determinant). By using the rules for expanding a determinant and working up the last column, which has only two non-zero elements, the patient reader should be able to arrive at the recurrence relation

$$D_N = d_N D_{N-2} - \alpha_{N-2} \beta_N D_{N-4} + \alpha_{N-4} \alpha_{N-2} \gamma_N D_{N-6} \tag{10}$$

for which we can use the starting conditions $D_{-2} = 1, D_0 = d_0, \gamma_2 = 0$ at $N = 2$. D_2, D_4 and so on can then be calculated for any assigned E value. If we use the estimates E_1 and $E_2 = E_1 + DE$, with DE small, then for each N the values of $D_N(E_1)$ and $D_N(E_2)$ will give a predicted energy E at which $D_N(E)$ would have been zero. We can find several roots for a particular N and we can also follow a particular root as N increases, to see whether it tends to a limit which is stable to some number of significant figures.

The recurrence relation (10) clearly can be handled by a program involving loops, with the α, d and D treated as arrays, and that is a standard way to deal with it. However, it is easy to produce an alternative version using names such as A, B, C, D for the four D_N which appear in the recurrence relation. I took the second (miserly RAM) view in some previous programs and show how to do the same thing here. I give first a list of the letters and their uses and then the program. The program works directly on a PC-1211 and can be modified for a TI-58 by simply using the alphabetic translation A = store 1, B = store 2, etc.

List of Symbols

A, B, C, D : G, H, I, J	Determinant values
E, F : Z	Trial energies, projected energy
L, M, P	λ, μ, β
N, Q	Order number, scaling factor
K, R, S, T, U, V	Various parameters

Program

```
 10 INPUT M, L, P : K = M − 4 * P * P
 20 INPUT N, E, F, : C = 0 : D = 0 : I = 0 : J = 0
 30 B = 1 : H = 1 : T = 8 * P : N = N − 2
 40 R = (4 * N + 2) * P − E : S = R + E − F
 50 N = N + 2 : U = K * N * (N − 1)
 55 R = R + T : S = S + T
 60 V = L * N * (N − 1) * (N − 2) * (N − 3)
 70 A = R * B + U * C + V * D
 80 G = S * H + U * I + V * J
 90 Z = E + (F − E)/(1 − G/A)
100 PRINT Z, N : Q = 1
110 IF ABS (A) > 1 E80 LET Q = 1E − 6
```

120 D = Q * C : C = Q * B : B = Q * A
130 J = Q * I : I = Q * H : H = Q * G
140 GOTO 50

Using N = 0 or 1 as input gives even or odd states. Line 110 prevents overflow: the value 1 E80 would be replaced by 1 E30 on microcomputers which overflow at 1 E38. Using $\beta = 2, \mu = 0, \lambda = 1$, I obtained the even eigenvalues 1.060 362 090 and 7.455 697 938, stable at $N \cong 60$. The odd eigenvalue 3.799 673 029 also becomes stable, at $N \cong 50$. With $\beta = 5$ the even eigenvalue 50.256 254 51 becomes stable at $N \cong 140$ and agrees with the accurate result of §7.4.

The procedure described above has been used for perturbed oscillator problems [6]; the recurrence relation approach is particularly simple when the matrix has only one non-zero element beyond the diagonal element in each row. A *tridiagonal* matrix is one with only the β, d and α elements non-zero in each row and is, of course, easy to handle by the method used above. Indeed, some of the techniques for diagonalising a general matrix proceed by first transforming it to tridiagonal form [7]. One way to accomplish a similar result is to deliberately choose the basis states sequentially so that the energy operator H automatically gives a tridiagonal matrix. This is the 'chain model' approach (in which the states resemble a linear chain with nearest-neighbour interactions) as discussed by Haydock [8]. In essence it involves using a basis formed from the functions $H^n \phi$, where ϕ is some starting or reference function. It should be clear from the above discussion that the series method (§7.4) and the Hill determinant method are alternative techniques for treating the same recurrence relations. The power series method literally works out the wavefunction at a specific x, whereas the matrix approach considers the global form of the wavefunction, as represented by the set of coefficients A_n. My own work indicates that the power series approach is better when applicable [9] and it has several obvious advantages. It can be used with Dirichlet or Neumann boundary conditions imposed at an arbitrary x value, and it will work even when the number of non-zero elements per row would render the recurrence relation for the determinants D_N very complicated. For example, in a relativistic calculation a small D^4 term might be included in the kinetic energy operator [10]. This gives two elements beyond the diagonal in each row of a matrix approach, complicating considerably the evaluation of the D_N, whereas it makes no extra difficulty for the power series approach.

As examples of methods which directly yield tridiagonal matrix problems I should mention the simple finite-difference approach to the one-dimensional Schrödinger equation (§10.2) and the use of cubic spline functions in a collocation approach to that equation [11]. Collocation methods use postulated wavefunctions which are linear combinations of some basis set ϕ_n, but concentrate on ensuring that the solution works at some selected set of points in space. As the number of sample points is increased we suppose, of course, that the wavefunction gets 'better' in some global sense. Although the series solution method is better for one-dimensional problems, the matrix eigenvalue approach can be used for more complicated problems; a remarkable example of the use of recurrence relations is the Pekeris [12] calculation of the helium atom ground state energy. Frost [13] discusses the recurrence relation method in a general context, stressing its value in avoiding the computation of integrals to get matrix elements. Some authors seem very concerned to produce a symmetric Hill matrix, since such a matrix can only give real eigenvalues, whereas a non-symmetric matrix (like that studied in my example) might give some complex ones. For some λ and β values it is easy to arrange that complex eigenvalues appear in my x^4 oscillator example; nevertheless, I find that proceeding to the D_N for sufficiently large N yields real limiting eigenvalues, so the search for a symmetric matrix might not be as crucial as some authors think. In any case, both for this problem (and for the solution of the equation $f(x) = 0$) we have a clear fail-safe 'principle of continuity'; if the function D (or f) has opposite signs at two E (or x) values, both real, then there must be at least one real root between those values.

8.4   Other types of eigenvalue calculation

In this short book I cannot give an exhaustive treatment of all the methods for eigenvalue calculation, but I note here a few which I think might prove capable of microcomputer implementation. (I willingly turn them over to any interested reader as a research project.)

If we start from the Schrödinger equation in the form

$$(H - E)\psi = 0 \tag{11}$$

and take N basis functions f_n, then all the inner products $\langle f_n | (H - E) | \psi \rangle$ must be zero for an exact (E, ψ) pair. If ψ is postulated to be a linear combination of N functions ϕ_n, then the requirement that the N inner products shall be zero leads to an $N \times N$ matrix eigenvalue problem. In the Rayleigh-Ritz approach the sets f and ϕ are identical, while in the Galerkin method they differ. Another approach, the *local energy method* [14], works out $\epsilon(\psi) = H\psi/\psi$ directly at many points of space, adjusting ψ until the local energy $\epsilon(\psi)$ has the smallest possible fluctuation as indicated by the standard deviation over the set of sample points. For an exact eigenfunction $\epsilon(\psi)$ is constant and equal to the eigenvalue at all points of space. The *finite element method* [15] is essentially a version of the Rayleigh-Ritz method in which the expectation values are calculated numerically as sums of contributions from a discrete set of volume elements which fill the space. It seems to me that there are still many hybrid methods which need investigation. For example, it doesn't seem to be essential that the number of f and ϕ functions should be the same in a Galerkin method, and it would be quite possible to look at the standard deviation of a set of quantities of form $\langle \phi_n | H | \psi \rangle / \langle \phi_n | \psi \rangle$ as a variant of the local energy method.

As an interesting point which links this chapter with chapter 3, I note that a study of the Brillouin-Wigner series for the energy (solution 3.12) shows that all eigenvalues are real for a real tridiagonal matrix such that $A_{jk}A_{kj}$ is positive for all j and k. Wilkinson [7] calls such matrices pseudo-symmetric.

Notes

1. R S Chasman 1961 *J. Math. Phys.* **2** 733
2. H S Green 1968 *Matrix Methods in Quantum Mechanics* (New York: Barnes and Noble)
3. J P Killingbeck 1975 *Techniques of Applied Quantum Mechanics* (London: Butterworths)
4. S T Epstein 1974 *The Variation Method in Quantum Chemistry* (New York: Academic Press)
5. M H Choudhury and D G Pitchers 1977 *J. Phys. B: At. Mol. Phys.* **10** 1209
6. S N Biswas, K Datta, R P Saxena, P K Srivastava and V S Varma 1973 *J. Math. Phys.* **14** 1190
7. J H Wilkinson 1965 *The Algebraic Eigenvalue Problem* (Oxford: Oxford University Press)
8. R Haydock 1980 *The Recursive Solution of the Schrödinger Equation* in *Solid State Physics* **35** 216 (New York: Academic Press)
9. J Killingbeck 1981 *Phys. Lett.* **84A** 95
10. M Znojil 1981 *Phys. Rev.* D **24** 903
11. B W Shore 1973 *J. Chem. Phys.* **58** 3855
12. C L Pekeris 1958 *Phys. Rev.* **112** 1649
13. A A Frost 1964 *J. Chem. Phys.* **41** 478
14. A A Frost, R E Kellogg and E C Curtis 1960 *Rev. Mod. Phys.* **32** 313
15. P M Prenter 1975 *Splines and Variational Methods* (New York: John Wiley)

9 Hypervirial-perturbation methods

9.1 Introduction

In the next few sections I give a brief sketch of some parts of Rayleigh–Schrödinger perturbation theory. I have written in detail about perturbation theory elsewhere [1, 2] and concentrate here on a few ideas which are directly useful in numerical work on a microcomputer. To guide the reader I list below the main themes which are strongly related to material in other sections of the book.

(1) The perturbation series for the perturbed oscillator and perturbed hydrogen atom are derived using a microcomputer hypervirial method in §9.7. The series are divergent and can be treated using Wynn's algorithm (§6.4) to form Padé approximants, or by a renormalisation trick which I describe.

(2) The simple formula for the first-order energy E_1 has two important applications. It is used to calculate expectation values without explicitly using the wavefunctions to do integrals (§9.4) and it is used to improve the accuracy of a simple method for finding energies by using finite differences (§10.3).

(3) The Hylleraas principle for E_2 is useful in connection with one of my detailed case studies, the theory of the quadratic Zeeman effect (§12.4). As a mathematical principle it is simply a disguised form of the iterative inverse calculation which is useful in matrix theory (§3.4) and in Padé approximant theory (§6.3).

9.2 Rayleigh–Schrödinger theory

If we start from the perturbed Schrödinger equation

$$(H_0 + \lambda V)\psi = E\psi \tag{1}$$

114

and compare it with the unperturbed equation

$$H_0 \phi_0 = E_0 \phi_0 \tag{2}$$

then, by taking the inner product of (1) with ϕ_0 and of (2) with ψ, we derive the energy shift formula

$$(E - E_0) \langle \psi | \phi_0 \rangle = \langle \psi | \lambda V | \phi_0 \rangle. \tag{3}$$

If the energy and wavefunction are postulated to be power series in λ, the strength of the perturbation, we can write the expansions

$$\psi = \phi_0 + \lambda \psi_1 + \lambda^2 \psi_2 + \ldots \tag{4}$$

$$E = E_0 + \lambda E_1 + \lambda^2 E_2 + \ldots . \tag{5}$$

Inserting these into (1) and taking the terms of first order in λ gives

$$(H_0 - E_0)\psi_1 = (E_1 - V)\phi_0 \tag{6}$$

where the first-order energy shift coefficient E_1 is given by

$$E_1 = \langle \phi_0 | V | \phi_0 \rangle \tag{7}$$

as can be seen from the energy shift formula or by taking the inner product of (6) with ϕ_0. If the perturbed wavefunction is chosen to obey *intermediate normalisation*, $\langle \psi | \phi_0 \rangle = 1$, then we have $\langle \psi_1 | \phi_0 \rangle = 0$, so that ψ_1 is orthogonal to the unperturbed wavefunction. From the equations (3), (4), (5) and (6) we can deduce the result

$$E_2 = \langle \psi_1 | V | \phi_0 \rangle = \langle \psi_1 | (E_0 - H_0) | \psi_1 \rangle \tag{8}$$

and by looking at the equation for the second-order function ψ_2 we can show that

$$E_3 = \langle \psi_1 | (V - E_1) | \psi_1 \rangle \tag{9}$$

so that knowledge of ψ_1 will give us E_2 and E_3.

For a perturbed oscillator or hydrogen atom it is often easy to calculate ψ_1 explicitly. For example, the hydrogen atom has the ground state wavefunction $\exp(-r)$ for which

$$[-\tfrac{1}{2}\nabla^2 - r^{-1}] \exp(-r) = -\tfrac{1}{2} \exp(-r) \tag{10}$$

A little algebra shows that for this ϕ_0

$$(H_0 - E_0)r^N \phi_0 = N[r^{N-1} - \tfrac{1}{2}(N + 1)r^{N-2}]\phi_0 \tag{11}$$

so that, for example,

$$(H_0 - E_0)r^2 \phi_0 = 2[r - \tfrac{3}{2}]\phi_0. \tag{12}$$

Comparing this with (6) shows that for the perturbation $V = \lambda r$ we have the first-order function

$$\psi_1(r) = \tfrac{1}{2}[3 - r^2]\phi_0. \tag{13}$$

For the reader who does not see this at once, I should point out that $\tfrac{3}{2}$ is E_1 i.e. $\langle \phi_0 | r | \phi_0 \rangle$. This must be so, since $(H_0 - E_0) r^2 \phi_0$ has to be orthogonal to ϕ_0 i.e. give zero inner product with ϕ_0. Also the 3 in ψ_1 is $\langle \phi_0 | r^2 | \phi_0 \rangle$ and is included to keep ψ_1 orthogonal to ϕ_0, while a -1 factor overall is used to convert $V - E_1$ to $E_1 - V$. From ψ_1 we can now calculate E_2 and E_3 as expectation values involving powers of r; the resulting ground state energy series is

$$E = -\tfrac{1}{2} + \tfrac{3}{2}\lambda - \tfrac{3}{2}\lambda^2 + \tfrac{27}{4}\lambda^3 \dots . \tag{14}$$

Higher order terms follow by solving for ψ_2, etc, but in §9.7 I shall show how to get them by using a microcomputer program which proceeds without needing to solve the equations for the ψ_n.

9.3   The Hylleraas principle for E_2

There are many problems for which ψ_1 cannot be found exactly but for which ψ_1 and E_2 can be estimated by using the Hylleraas functional

$$F(f) = 2\langle f | V | \phi_0 \rangle + \langle f | (H_0 - E_0) | f \rangle. \tag{15}$$

where f is a trial function. This functional relates to the iterative procedure for inverses (§3.4, exercise 6.3), since another way to write E_2 is as $-\langle g | (H_0 - E_0)^{-1} | g \rangle$ with $g = (V - E_1)\phi_0$. If the trial function is of form $f = \psi_1 + \chi$, then a little algebra using (6) and (15) shows that

$$F(f) = E_2 + \langle \chi | (H_0 - E_0) | \chi \rangle. \tag{16}$$

This means that F has a stationary value for $f = \psi_1$. For the ground state the stationary value is a minimum and E_2 has a negative value. By varying f until $F(f)$ is minimised we thus get an upper bound to the ground state E_2 value. From the optimum f, the 'best' estimate of ψ_1, an approximate E_3 value can also be found. The best known problem for which the Hylleraas principle has been used is the two-electron ion [3] with Hamiltonian

$$-\tfrac{1}{2}[\nabla_1^2 + \nabla_2^2] - Z(r_1^{-1} + r_2^{-1}) + r_{12}^{-1}. \tag{17}$$

The inter-electron repulsion term r_{12}^{-1} is the perturbation and the energy series takes the form $-Z^2[E_0 + E_1 Z^{-1} + E_2 Z^{-2} \dots]$. The inverse atomic number Z^{-1} thus measures the strength of the perturbation. The Z^{-1} energy series actually converges if Z is sufficiently large, and there is an enormous literature connected with it. My review article [2] gives more details and references. A very recent application of Z^{-1} perturbation theory is the work of Knight [4] on four-electron systems.

9.4 The expectation value problem

The power series method of §7.4 gives very good energies for polynomial potentials and the finite-difference method of §10.2 is also capable of giving good energy values. However, to find an expectation value such as $\langle x^2 \rangle$ for a bound state it might appear that we need to have the eigenfunction ψ for all x, so that we can integrate $\psi^2 x^2$ over all x. This is *not* so; to show this I invoke the formula for the first-order energy shift E_1 in perturbation theory. If the energy is calculated with a small term $\pm \mu x^2$ included in the potential, then the resulting perturbed energies for the Hamiltonians $H \pm \mu x^2$ will be

$$E(H \pm \mu x^2) = E_0 \pm \mu E_1 + \mu^2 E_2 \pm \ldots \tag{18}$$

with μ sufficiently small we conclude that

$$\langle \phi_0 | x^2 | \phi_0 \rangle$$

$$= \frac{1}{2\mu} [E(H + \mu x^2) - E(H - \mu x^2)]. \tag{19}$$

What I am doing here is a kind of numerical differentiation, and it is possible to improve the result for $\langle x^2 \rangle$ even further by computing $E(H \pm 2\mu x^2)$ and using a Richardson extrapolation (§4.3). The central point is this; for the one-dimensional Schrödinger equation the calculation of expectation values can in principle be reduced to a sequence of energy calculations, and the energy calculations can be done to high accuracy. I have found it to be one of the most interesting intellectual puzzles in my microcomputer work to discover how to convert various calculations into a form which involves only eigenvalue computation, since the latter can be done fairly easily. As the reader will see, here is another case where the analysis has to be done carefully before the computing begins!

To illustrate the idea I use the Schrödinger equation

$$(-D^2 + \mu x^2 + \lambda x^4)\psi = E\psi \tag{20}$$

for which I have already shown how to get energy values by the power series approach (§7.4) and by the Hill determinant approach (§8.3). For the ground state the value of E with $\mu = 0$ and $\lambda = 1$ is $1.060\,362\,09$. When a small term μx^2, with $\mu \simeq 0.01$, is added to the potential the energy does not change very much, so we are locked on to a good starting estimate for E already. This cuts down the computing time required considerably, so that the longest part of the calculation is the initial ($\mu = 0$) energy determination. Using the series method I found the following results at $\lambda = 1$.

μ	E	$\langle x^2 \rangle$
0	1.060 362 09	
0.01	1.067 588 78	
−0.01	1.053 107 79	0.362 0248
0.02	1.074 788 11	
−0.02	1.045 825 63	0.362 0310

Doing a Richardson extrapolation on the $\langle x^2 \rangle$ values and carefully conceding that the energies might be in error by $\pm\frac{1}{2}$ in the last digit, I find the estimate $\langle x^2 \rangle = 0.362\,0227$ (± 3). To the number of digits quoted this agrees with the accurate result of Tipping [5]; his paper gives many results for the quartic oscillator and contains some beautiful applications of hypervirial relations and sum rules.

To find expectation values such as $\langle x^4 \rangle$, $\langle x^6 \rangle$, etc for the above example with the potential function x^4 we do not need to use more eigenvalue calculations, but can use hypervirial relations (§9.6). For this problem the following hypervirial relation holds (for even N);

$$(N + 1)E\langle x^N \rangle = (N + 3)\langle x^{N+4} \rangle - \tfrac{1}{4}N(N^2 - 1)\langle x^{N-2} \rangle. \tag{21}$$

The choice $N = 0$ gives $3\langle x^4 \rangle = E$, while the choice $N = 2$ gives

$$10\langle x^6 \rangle = 6E\langle x^2 \rangle + 3. \tag{22}$$

From the results above I find $\langle x^4 \rangle = 0.353\,454\,030$ (± 2) and $\langle x^6 \rangle = 0.530\,3251$ (± 2) and from knowledge of E and $\langle x^2 \rangle$ there follow values for the higher $\langle x^N \rangle$. Similar procedures work for any problem with a polynomial potential function; I treat the quadratic Zeeman effect in §12.4.

9.5   Calculating $\psi(0)$ for radial problems

The intellectual puzzle of how to reduce everything to an energy calculation gets really interesting if we want to calculate a local quantity, e.g. the value of the (normalised) wavefunction at the origin. In a recent paper [6] various uses for this quantity are cited and a method of calculating it by using finite-difference methods is explained. However, for s states of a radial Schrödinger equation of the form

$$-\tfrac{1}{2}D^2R + \sum_{-1}^{N} V(n)r^nR = ER \tag{23}$$

the power series method can be used with more ease than the finite-difference method. What matters is the general procedure for finding the gradient $DR(0)$ at $r = 0$. Since the full wavefunction $\psi(r)$ is equal to the product $r^{-1}R(r)$ it follows that $DR(0)$ equals $\psi(0)$. Multiplying the radial equation by DR and integrating from 0 to ∞ produces the following result (with $V(r)$ denoting the potential function);

$$[DR(0)]^2 = 2 \int_0^\infty R^2 (DV) \, dr. \qquad (24)$$

(To get the result we integrate $VRDR$ by parts and remember that R is zero at $r = 0$ and $r = \infty$.) This result shows that $\psi^2(0)$ is equal to the expectation value of the gradient DV of the potential function. This quantity can be calculated using energy computations, as outlined in the previous section, so we have reduced the calculation of a local quantity $\psi^2(0)$ to energy computations! I thought for some time about how to extend this argument to calculate $\psi^2(r)$ for any r in terms of energy calculations. Only recently did I see how it can be done. The trick is to see that a perturbing delta function potential $\lambda\delta(r - R)$ at $r = R$ would give a first-order energy shift equal to the expectation value of the perturbation. This will be $\lambda\psi^2(r)$. So far so good, but how can delta-simulating potential be put into the calculation? Well, my current view is that an explicit potential needn't be used; it is sufficient to make the calculation proceed *as though* the wavefunction were constrained to suffer a discontinuity of slope of magnitude λ at $r = R$. This can be done in some eigenvalue calculational techniques, particularly finite-difference ones, and preliminary calculations suggest that this approach gives accurate results.

9.6 Hypervirial relations

In the Heisenberg picture of quantum mechanics the state vectors are fixed while an operator A has a rate of change proportional to the commutator $[H, A]$, H being the energy operator. The analogous result in the Schrödinger picture is that the rate of change of the *expectation value* $\langle A \rangle$ is proportional to the expectation value $\langle [H, A] \rangle$. For a stationary state then, with time-independent expectation values, any commutator $[H, A]$ has zero expectation value. For a one-dimensional problem with the (time-independent) Schrödinger equation

$$-\alpha D^2 \psi + V\psi = E\psi \qquad (25)$$

we require that $\langle \psi | [H, A] | \psi \rangle$ shall vanish for any A. If A takes the product form $A = fD$, with f a smooth function of x, then the commutator $[H, A]$ can be explicitly evaluated and set equal to zero. A little algebra shows that

$$[H, fD]\psi = -\alpha f''\psi' - 2\alpha f'\psi'' - fV'\psi. \tag{26}$$

On replacing $-\alpha\psi''$ by $(E - V)\psi$ from the Schrödinger equation and integrating by parts to get rid of the ψ' term, we work out the expectation value $\langle\psi|[H, fD]|\psi\rangle$. Setting this equal to zero yields the relation

$$2E\langle f'\rangle = \langle fV' + 2f'V\rangle - \tfrac{1}{2}\alpha\langle f'''\rangle. \tag{27}$$

The special case $f = x$ gives

$$2\langle T\rangle = \langle xV'\rangle \tag{28}$$

if we use the symbol T for the kinetic energy operator. This result is called the *virial theorem*; it also holds for classical bound state motions if time averages are used instead of quantum mechanical expectation values. *Hypervirial* relations are obtained by making other choices of f; they also have classical analogues [6]. The choice $f = x^{N+1}$ with the potential $V = \Sigma V_n x^n$ gives the result

$$2E(N + 1)\langle x^N\rangle = \Sigma V_n(2N + 2 + n)\langle x^{N+n}\rangle$$
$$-\tfrac{1}{2}\alpha N(N^2 - 1)\langle x^{N-2}\rangle. \tag{29}$$

This formula has an obvious use; if E and a sufficient number of the $\langle x^N\rangle$ are known (analytically or numerically) then it allows computation of other $\langle x^N\rangle$ values. For example in §9.4 I noted that, knowing E and $\langle x^2\rangle$ for the case $V = x^4$, we can find $\langle x^4\rangle$, $\langle x^6\rangle$, etc.

As a simple example of the use of the virial theorem in *classical* mechanics I use the anharmonic oscillator [7] with the equation of motion

$$\ddot{x} = -x - \lambda x^3. \tag{30}$$

The kinetic energy is $(\dot{x}^2/2)$ and the potential energy is $(x^2/2) + \lambda(x^4/4)$. I now propose that the motion takes the approximate form $x = A \cos Wt$, with W to be determined. (It doesn't take this *exact* form, of course, but the principle is the same as that of using a trial wavefunction in a quantum mechanical variational calculation.) The virial theorem (28) says that

$$\langle\dot{x}^2\rangle = \langle x^2 + \lambda x^4\rangle. \tag{31}$$

Working out the time averages over one cycle produces the result

$$W = 1 + \tfrac{3}{8}\lambda A^2 + \dots \tag{32}$$

for small amplitude A. This agrees to order λA^2 with the result obtained by using the exact integration method of §5.7.

9.7   Renormalised perturbation series

I return to the perturbed oscillator problem

$$-D^2 \psi + x^2 \psi + \lambda x^4 \psi = E\psi \tag{33}$$

to explain the use of hypervirial relations in calculating perturbation series. If this Schrödinger equation is attacked by the power series method (§7.4) then a factor $\exp(-\beta x^2)$ is built into the postulated form for ψ. Such a factor actually represents the exact ground state function for a harmonic oscillator with a potential term $4\beta^2 x^2$. A similar approach can be used in the perturbation problem; we take the unperturbed system to be a renormalised oscillator with an adjustable x^2 coefficient. To make sure that the actual problem treated is the same as the original one we write the potential as

$$V = \mu x^2 + \lambda(x^4 - Kx^2) \tag{34}$$

but make sure to set $\mu = 1 + \lambda K$ in any numerical work. This ensures that the coefficient of x^2 is 1. However, in perturbation theory, where terms are collected according to powers of λ, the results obtained from the perturbation series *do* vary with K, even though the *exact* eigenvalues are independent of K. I shall present some illustrative results later, but first outline the formal procedure.

The first step is to insert the series expansions

$$E = \Sigma E(n)\lambda^n : \langle x^m \rangle = \Sigma A(m, n)\lambda^n \tag{35}$$

into the hypervirial relation (29), with $V_2 = (\mu - \lambda K)$, $V_4 = \lambda$ and all other V_n zero. Extracting the coefficients of λ^M from the resulting equation gives a recurrence relation:

$$(2N + 2)\sum_{0}^{M} E(k)A(N, M - k)$$

$$= \mu(2N + 4)A(N + 2, M) - K(2N + 4)A(N + 2, M - 1)$$

$$+ (2N + 6)A(N + 4, M - 1) - \tfrac{1}{2}N(N^2 - 1)A(N - 2, M). \tag{36}$$

To complete the scheme we need a relation between the Es and the As. We get this by supposing that λ varies slightly. The energy change will be $\langle x^4 - Kx^2 \rangle \delta\lambda$ from the first-order energy formula (§9.2), but it will also be $\delta\lambda$ times the derivative of the energy series. Comparing coefficients then gives the result

$$(n + 1)E(n + 1) = A(4, n) - KA(2, n). \tag{37}$$

The astonishing fact is that the equations (36) and (37) suffice to calculate the full set of E and A coefficients. All that is needed is the value of E_0, the unperturbed energy, which is $(2N + 1)\mu^{1/2}$ for this case. The input for the calculation is λ and K. μ and E_0 are worked out by the program, and any desired quantity can be printed out. In my own work [8] I looked at the way in which the partial sums of the series for the energy and for $\langle x^2 \rangle$ could be made to give good numerical results for those quantities. At $K = 0$ the perturbation series are the conventional Rayleigh-Schrödinger ones, which diverge quickly and so do not give satisfactory numerical results. As an example I show some sums of the ground state energy series for the case $\lambda = 1$, with $K = 0$ and $K = 4$.

N	$K = 0$	$K = 4$
6	-1800	1.392 342
7	189 77	1.392 348
8	$-226\,712$	1.392 351
9	302 93 10	1.392 352
10	$-447\,811\,21$	1.392 351

The accurate energy for this case is 1.392 351 64 and it is clear that the use of the K parameter 'tames' the usually divergent series quite remarkably. By varying K carefully the results can be made even more accurate and the method even works for the notorious double well potential with $V = -x^2 + \lambda x^4$ [8]. There is, of course, another way to deal with the divergent $K = 0$ series, namely to sum it by using Padé approximants, e.g. by using the Wynn algorithm (§6.4). However, the renormalised series trick seems to give the results to greater accuracy and more easily. Further, the hypervirial approach also gives the series (and their sums) for quantities such as $\langle x^2 \rangle$, $\langle x^4 \rangle$, etc, and it does so without explicitly calculating the perturbed wavefunctions ψ_1, ψ_2, etc, which would be required in the orthodox approach (§9.2). This is really a remarkable illustration of my remark (§2.5) that analysis can often help to cut down the number of quantities which have to be calculated.

Well, after the publicity, what about the program? Here it is, in Pet style, as I ran it on a 2001 series Pet (with an old ROM, but with a RAM extended to 32K). The results beyond $N = 10$ are, of course, obtainable, but show slight rounding error in the later digits which makes it difficult to get any greater accuracy.

```
 5 DIM B(26, 26), E(26)
10 INPUT K, L : D = L
15 B(2, 1) = 1 : MU = 1 + K * L : E(0) = SQR(MU)
20 E = E(0) : FOR M = 0 TO 26
25 FOR N = 0 TO 20 STEP 2
30 S = 0 : FOR P = 0 TO M
40 S = S + E(P) * B(N + 2, M + 1 − P) : NEXT P
50 S = S * (2N + 2) + 0.5 * N * (N * N − 1) * B(N, M+1)
55 T = MU * B(N+4, M+1) − K * B(N+4, M)
60 S = S − (2 * N+4) * T − (2 * N+6) * B(N+6, M)
```

```
 65  B(N+4, M+1) = (S/MU)/(2 * N+4) : NEXT N
 75  Z = B(6, M+1) − K * B(4, M+1)
 80  E(M+1) = Z/(M+1) : E = E + E (M+1) * D
 90  D = D * L : PRINT E (M+1), E
100  IF M = 9 THEN END : NEXT M
```

The program makes an essential use of arrays, but since the array sub-scripts on the Pet cannot be negative the subscripts have to be changed from their values in the theoretical equations. I used the change of variable $A(N, M) \to B(N + 2, M + 1)$ to move the subscripts so that they are all non-negative. In line 15 the quoted value of $E(0)$ is for the ground state calculation. To deal with excited states a factor $(2N + 1)$, with some integer N, should be included. The energy series at $K = 0$ is an alternating one, whereas the series at $K = 4$ (for $\lambda = 1$) is not; the terms are in bunches of plus or minus type. Because of this I did not try to do a Padé analysis on the renormalised series, since almost all of the effective uses of Padé approximants in the literature have involved alternating series. I guessed that using the approximants would not improve the results further. I guessed wrong: Dr Elizabeth Austin has done careful double precision calculations which show that Padé approximants based on the renormalised series can give energies which are as accurate as those obtainable by any other technique. I should perhaps emphasise something at this stage; the kind of work described in this section is valuable for the light which it throws on the ways in which divergent perturbation series can be formally calculated and then 'tamed' to give numerical results. Several authors have gone beyond this and have advocated the use of this approach as a *practical* means of eigenvalue calculation. In fact it becomes very tedious if the pertur-bation is strong or if excited states are treated. I have no hesitation in declaring that the power series or finite-difference methods are the speedy practical methods if all we want is to get good eigenvalues for a one-dimensional Schrödinger equation.

For the perturbed hydrogen atom system with the kinetic energy operator $-\frac{1}{2}D^2$ and the potential energy

$$-r^{-1} + \lambda r = -\mu r^{-1} + \lambda [r - Kr^{-1}] \tag{38}$$

(for s states) the choice $\mu = 1 - K\lambda$ for the parameter μ gives a hypervirial calculation analogous to that for the perturbed oscillator. My BASIC program for this problem is given in Appendix 2 and my paper [8] discusses various ways in which the programs can be used. In particular I test the idea that the parameter K should be chosen to make $(\partial E/\partial K)$ (as given by the series) zero, since the exact energies are certainly independent of the dummy parameter as are the $\langle r^N \rangle$. While the work of this section is concerned with Rayleigh–Schrödinger series which diverge (at $K = 0$) I should point out that in §7.3 I gave an example for which the energy series converges (indeed terminates) but does not give the correct perturbed energy. Perturbation theory is an endless source of examples for those interested in the eccentric behaviour of power series!

9.8   The sum-over-states formalism

Most quantum mechanics textbooks use the sum-over-states formalism. The idea is to represent the first-order perturbed function ψ_1 as a linear combination of the eigenfunctions ϕ_n of the unperturbed Hamiltonian H_0. These basis functions thus have the property $H_0\phi_n = E_n\phi_n$. By direct substitution the reader may verify that the following form of ψ_1 satisfies equation (6);

$$\psi_1 = \sum_{n \neq 0} \phi_n \langle n|V|0 \rangle (E_0 - E_n)^{-1}. \tag{39}$$

This way of expressing ψ_1 is sometimes useful, although it is often the case that there is a strong contribution to ψ_1 from basis functions which are continuum functions and not bound state functions (see §12.5 for an example). Estimating E_2 and ψ_1 by using the Hylleraas principle (§9.3) avoids any explicit mention of the continuum solutions for the unperturbed Hamiltonian. If E_2, as given by equation (8), is expressed in sum-over-states form it becomes

$$E_2 = \sum_{n \neq 0} |\langle 0|V|n \rangle|^2 (E_0 - E_n)^{-1}. \tag{40}$$

In Hartree-Fock theory the wavefunction ψ of a many-electron system is taken as an antisymmetrised function based on a simple product function in which each electron is assigned to a single-particle orbital. Choosing the occupied orbitals to minimise the energy expectation value gives the Hartree-Fock orbitals. As I have explained elsewhere [1] formula (39) can be used to explain why the Hartree-Fock ψ gives good values for one-electron expectation values. Formulae (39) and (40) also can be used to explain why in some atoms the traditional Hund's rules are not obeyed [9, 10]. In many-body theory, then, the sum-over-states formalism can be useful for giving a semi-quantitative account of various important effects, and the so-called diagrammatic perturbation theory used in atomic theory leans heavily on that formalism. For the kind of one-particle examples which I treat in this book, however, there is no doubt that the sum-over-states approach is much more cumbersome than the approach of §9.2 and 9.3. If I may drive home the point: would you rather calculate using, say, the function $x - x^2$ or using that function expressed as an infinite Fourier expansion? (Think about it for a microsecond or so!)

Exercises

1. In §5.7 I introduced the WKB formula for the energy

$$(n + \tfrac{1}{2})\pi = \int_{X_1}^{X_2} [E - V(x)]^{1/2} \, dx. \tag{41}$$

Here X_2 and X_1 are the classical turning points, at which the integrand is zero. To find the expectation value $\langle U(x) \rangle$ for some function $U(x)$ we should (according to §9.4) add a tiny term $\mu U(x)$ to the potential $V(x)$ and compute the quantity $\partial E / \partial \mu$. In this case the rate of change of the left-hand side with respect to μ is zero, and this enables us to obtain an expression for $\partial E / \partial \mu$. What is it?

2. Consider an oscillator in a box,

$$-D^2 \psi + x^2 \psi = E\psi \tag{42}$$

with $\psi(0) = \psi(1) = 0$. Find the eigenvalues and eigenfunctions if the x^2 potential is neglected and then calculate the first-order energy shift (§9.2) when it is included as a perturbation.

Solutions

1. One of the terms in the derivative vanishes because the integrand is zero at the limits of integration. The result is

$$\langle U(x) \rangle = \frac{\displaystyle\int_{X_1}^{X_2} U(x)[E - V(x)]^{-1/2} \, dx}{\displaystyle\int_{X_1}^{X_2} [E - V(x)]^{-1/2} \, dx}. \tag{43}$$

This looks like a classical time average, since the classical speed v is proportional to $[E - V]^{1/2}$ at each point and so v^{-1} is a measure of the 'fraction of time' spent in a particular region during the classical motion. For the excited states, where the WKB formula becomes accurate, the expectation values accordingly are well given by the classical time averages.

2. The normalised eigenfunctions are of form

$$\phi_n = \sqrt{2} \sin (n\pi x) \tag{44}$$

and the eigenvalues are $E_n = n^2 \pi^2$. The integral for the expectation value of x^2 is of the type encountered in Fourier analysis and the energy to first order is

$$E_n = n^2 \pi^2 + \left(\frac{1}{3} - \frac{1}{2n^2 \pi^2} \right).$$ (45)

Accurate eigenvalues can be found using the series method of §7.4. The formula gives good results for excited states, since the gaps between the unperturbed states increase with n. Even for the ground state the estimate (10.1523) is not far from the correct value (10.1512).

Notes

1. J Killingbeck 1975 *Techniques of Applied Quantum Mechanics* (London: Butterworths)
2. J Killingbeck 1977 *Rep. Prog. Phys.* **9** 963
3. C W Scherr and R E Knight 1963 *Rev. Mod. Phys.* **35** 436
4. R E Knight 1982 *Phys. Rev.* **25A** 55
5. R H Tipping 1976 *J. Mol. Spectrosc.* **59** 8
6. J Killingbeck and S Galicia 1980 *J. Phys. A: Math. Gen.* **13** 3419
7. J Killingbeck 1970 *Am. J. Phys.* **38** 590
8. J Killingbeck 1981 *J. Phys. A: Math. Gen.* **14** 1005
9. R N Zare 1966 *J. Chem. Phys.* **45** 1966
10. H Odabasi 1969 *J. Opt. Soc. Am.* **59** 583

10 Finite-difference eigenvalue calculations

10.1 Introduction

In this chapter I deal with some simple finite-difference methods for calculating energy eigenvalues of the one-particle Schrödinger equation. The particular methods which I use are mainly my own, although they necessarily have much in common with any other finite-difference methods. My main aim has been to produce methods which give high accuracy but require only a small amount of microcomputer RAM. The use of a repetitive loop takes care of the space problem, while the use of the F function and of Richardson extrapolation (§10.2) produces high accuracy. The fundamental idea of the methods of this chapter, and also of the series method of chapter 7, is to classify the bound state wavefunctions according to the number of nodes which they have. This classification is not affected by the inclusion of any kind of nodeless multiplying factor in the wavefunction. The radial equation is treated in §10.3 by a refinement of a method which I invented several years ago for use with a simple type of programmable calculator. I have added a node counter and an energy interpolation routine to the program to make it appropriate for the kind of microcomputer which I deal with in this book. §10.3 contains what I regard as a novel and beautiful application of the simple perturbation theory which I outlined in chapter 9. The finite-difference methods developed in this chapter are applied to various problems in chapters 11 and 12.

10.2 The one-dimensional equation

The finite-difference quantity

$$\delta^2 \psi(x) = \psi(x + h) + \psi(x - h) - 2\psi(x) \tag{1}$$

can be written as a power series (§4.3)

$$\delta^2 \psi(x) = h^2 D^2 \psi(x) + \tfrac{1}{12} h^4 D^4 \psi(x) + \dots \quad (2)$$

If $\psi(x)$ is the wavefunction which obeys the Schrödinger equation

$$-\alpha D^2 \psi + V\psi = E\psi \quad (3)$$

then we can set

$$h^{-2} \delta^2 \psi(x) = D^2 \psi(x) + \tfrac{1}{12} h^2 D^4 \psi(x) + \dots$$

$$= \alpha^{-1}[V(x) - E]\psi(x) + VP \quad (4)$$

where the perturbation VP has a leading term of order h^2. This equation is the basis for my discussion of various finite-difference calculations of eigenvalues.

The most simple procedure is to ignore VP as a first approximation, and so study the equation

$$\delta^2 \psi(x) = \alpha^{-1} h^2 [V(x) - E]\psi(x). \quad (5)$$

This equation involves $\psi(x)$ and $\psi(x \pm h)$, so if we regard the quantities $\psi(Nh)$ as a column matrix, with $N = 0, 1, 2$, etc, we get a tridiagonal symmetric matrix eigenvalue problem (§8.3). This matrix interpretation of the problem, with the eigenvalues giving the energy values E_n, is quite a useful approach [1], but I wish to describe a procedure which is more akin to the recurrence relation method of §7.4. To simplify matters I will suppose the potential $V(x)$ to be an even function of x, so that the eigenfunctions can be classified as being even or odd. The value of $\psi(0)$ for an even state can be set at 1, while for an odd state $\psi(0)$ will be zero. Combining equations (1) and (5) gives $\psi(x + h)$ in terms of the two preceding values $\psi(x)$ and $\psi(x - h)$. In my early work on this subject I used a simple programmable calculator and to cut down the number of stores required I got into the habit of using *ratios* of the wavefunction at different points. For example, I used two quantities $R(x)$ and $F(x)$ which are defined as follows:

$$\psi(x + h) = \psi(x)R(x) \quad (6)$$

$$= \psi(x)[1 + h^2 F(x)]. \quad (7)$$

If these quantities are inserted into (5) the following results are obtained (as the reader can quickly verify)

$$R(x) + [1/R(x - h)] = 2 + \alpha^{-1} h^2 [V(x) - E] \quad (8)$$

$$F(x) - [F(x - h)/R(x - h)] = \alpha^{-1}[V(x) - E]. \quad (9)$$

My reason for using the F variable is that for small h the ratio R is close to 1. With $h = 0.01$, for example, an F of 5.231 corresponds to an R of 1.000 5231. On a machine which works to a fixed number of significant figures it is clear that using the F variable will give greater accuracy; the digit 1 at the start of the R variable is clearly the villain of the piece. As h tends to zero the quantity F tends to $h^{-1}\psi^{-1}D\psi$ i.e. is proportional to the so-called logarithmic derivative, the quantity which is required to be continuous in simple barrier penetration problems of quantum mechanics. To apply (8) or (9) we need to have some initial value for R or F and can then calculate successive R or F values along the x-axis, with some test energy E. In my early work I examined R or F directly to see whether E was a good estimate of an eigenvalue, but what I recommend nowadays is a procedure analogous to that used in the power series method (§7.4). The wavefunction ψ is calculated, using equation (8) or equation (9) for *two* trial energies E_1, E_2, on the same run of the program. If an eigenfunction with N nodes is being sought a node counter statement in the program 'turns on' the calculation of the projected energy

$$EP = E_1 + (E_2 - E_1)/[1 - \psi_2/\psi_1] \tag{10}$$

only after ψ_2 has had N nodes. (I suppose that $E_2 > E_1$, so that ψ_2 has its nodes earlier than ψ_1.) EP is actually a function of x; it is the interpolated energy which would have given $\psi(x) = 0$. As x increases, however, EP settles down to a limiting value, provided that $E_2 - E_1$ is not too large. This limiting energy corresponds to the boundary condition $\psi(\infty) = 0$. By using it as the E_1 value on the next run we quickly home in on an eigenvalue. The procedure is similar to that for the series method, but there are some differences. First (the good news) $V(x)$ is given as a function and need not be in the form of a polynomial or power series. Second (the bad news) the energies refer to the equation (5), which is *not* the exact Schrödinger equation. Noting that the term VP in (4) is a series in h^2 we can conclude from the basic perturbation theory of §9.2 that the true energy is related to the calculated energy by a formula of type

$$E(h) = E_0 + E_2 h^2 + E_4 h^4 + \dots \tag{11}$$

Here $E(h)$ is the energy calculated using stripwidth h and E_0 is the exact energy (for $h \to 0$). This type of behaviour is of the same kind as that encountered in Romberg integration (§5.4), so that we can simply 'plug in' to the methods of that section. For example, to get a simple improved estimate of E_0 we can form the quantity $[4E(h) - E(2h)]/3$, and so on, in a Richardson extrapolation process. This requires a few calculations, but the first one provides a good starting estimate which makes the later ones easier to perform.

What about the starting conditions? Well, if ψ is to be odd we must have $R(0) = \infty$, which we can accomplish on a microcomputer by setting $R(0) = 1E30$, say, or by formally setting $1/R(0)$ equal to 0, depending on how we write the program. $\psi(h)$ can be given some small starting value, for example h, which corresponds to having unit slope at the origin. For even states $\psi(-h) = \psi(h)$, which means that $R(0) = 1/R(-h)$. Putting this into (8) gives a value for $R(0)$ to start off the calculation. Another way to derive these starting conditions is to work out the quantity $\psi(h)$ as a Taylor series up to the h^2 term, noting that a smooth ψ function with even parity will have zero slope at $x = 0$.

As the reader will probably have gathered, over the last few years I have done many tests to check that using the F equation gives more accurate results than using the R equation. I have also checked that the Richardson extrapolation procedure applied to the $E(h)$ values gives accurate energies. In principle, of course, the series for $E(h)$ in powers of h is an asymptotic one, just as the Euler–Maclaurin series is (§5.3). However, the residual error term can be fairly easily rendered negligible on a microcomputer by taking h values which are not terribly small; h is of order 0.01 in most of my examples in this chapter. In a recent short paper [2] I have outlined how my approach relates to others in the literature. I should point out particularly that the use of the projected energy quantity $EP(x)$ lets the calculation 'find its own infinity', so that the operator does not have to use an artificial boundary at some x which he hopes is large enough to simulate $x = \infty$. The power series method (§7.4) *does* require such a procedure, although it doesn't involve any Richardson extrapolation to produce the accurate energy value.

I show below a simple program which will perform the calculation of $E(h)$. One thing in particular is noteworthy. Although the wavefunction appears in the theory and in the program, it need not be kept as an array. Only the ψ value at the current x value plays a role in the calculation; all the previous ψ values can be discarded without affecting the calculation. This feature is common to methods which find eigenvalues by looking for nodes at which ψ vanishes, and is one reason why my simple methods for microcomputers are of this type. For example, any convergence factor such as $\exp(-\beta x^N)$ built into the wavefunction does not change the nodal properties of ψ but can help to simplify the calculation, as I showed for the power series method in §7.4. When the absolute magnitude of ψ is being varied by such devices, it is wise to include an overflow control procedure to scale down ψ if it threatens to overflow. This is easier to do than for the power series method; only the current ψ needs to be scaled, rather than four or five numbers which are needed in the next application of the recurrence relation. I have found in my own work that it is useful to have the microcomputer compute (and sometimes print out) the number of 'nodes so far' for ψ_1 and ψ_2 (with associated energies E_1 and E_2). For example, if an eigenfunction with three nodes is being sought, the best situation is for ψ_1 to have two nodes and ψ_2 to have three, so that $E_2 > EP > E_1$. If $E_2 - E_1$ is large the evidence of the node counter is the one to rely on; the calculated EP in some such cases does not correctly fall between E_1 and E_2! However, as the eigenvalue is approached and $E_2 - E_1$ is made smaller the predictions of the node count and of the EP calculation become consistent. The node count is particularly useful when the operator has only a very hazy idea of where to look for the eigenvalue. It allows a direct search to be made for the Nth excited state (with N nodes); that the number of nodes increases with energy is a result of the Sturm-Liouville theory for second-order equations and the Schrödinger equation is usually of Sturm-Liouville type (see also exercise 1). The power series method of §7.4, when applicable, also lets us find excited state energies directly, but we don't always know which one we have, since ψ is only calculated at one position and not at all positions along the axis.

The program is set out below (in Pet style). Lines 35 to 55 involve a bit of clever manipulation to put in the correct starting values of F, R and x for even parity $(P = 0)$ or odd parity $(P = 1)$. Q is the number of nodes (e.g. 1 for the ground state of each parity type, 2 for the first excited state, and so on). By setting $K = 0$ in line 15 we get the simple method described above; $K = h^2/12$ gives the improved method of the next section, so I shall give numerical results only after I have described the improved method.

```
10 INPUT P, Q, H : H2 = H * H
15 A = 1 : K = H2/12 : D = 1
20 DEF FNA(X) = (X * X+1) * X * X
30 INPUT E, DE : PRINT : PRINT
35 IF P = 0 THEN 50
40 F1 = 1E30 : F2 = F1 : GOTO 55
50 W1 = 1 : W2 = 1 : R1 = 1E30 : R2 = R1
52 D = 2 : N = −1 : GOTO 60
55 R1 = 1 + H2 * F1 : R2 = 1 + H2 * F2 : W1 = 1 : W2 = 1
60 N = N + 1 : X = N * H : M = 1
70 G1 = (FNA(X) − E)/A : G2 = G1 − DE/A
80 F1 = F1/R1 + G1 : F2 = F2/R2 + G2
85 F1 = (F1 + K * G1 * G1)/D : F2 = (F2 + K * G2 * G2)/D
90 R1 = 1 + H2 * F1 : R2 = 1 + H2 * F2 : D = 1
100 IF ABS (W2) > 1E30 THEN M = 1E − 6
110 W1 = W1 * R1 * M : W2 = W2 * R2 * M
120 Q1 = Q1 + (1 − SGN (R1))/2 : Q2 = Q2 + (1 − SGN (R2))/2
130 IF Q > Q2 THEN 60
140 EP = E + DE/(1 − W2/W1)
150 PRINT "∧∧" EP : PRINT Q1;Q2 : GOTO 60
```

("∧∧" means move print line up two lines.)

10.3   A perturbation approach

I now return to the equation (4) and look at the question of how to deal with the *VP* term. In a numerical calculation the Richardson extrapolation allows for *VP* numerically, of course, but it would be useful to do part of the work directly in the numerical integration. In the evaluation of integrals, for example, using Simpson's rule instead of the trapezoidal rule makes the error of order h^4 instead of h^2, and so needs one less stage in a Romberg integration scheme (§5.4).

My own approach to this problem is to go back to perturbation theory (§9.2) and to ask what the first-order energy shift would be if a perturbing term $\frac{1}{12}h^2 D^4$ were added to a Hamiltonian. This shift would be the expectation value

$$E_1 = \tfrac{1}{12}h^2 \int \psi D^4 \psi \, dx. \tag{12}$$

To lowest order, then, this integral gives the energy shift causes by using $-h^{-2}\delta^2 \psi$ instead of the correct kinetic energy term $-D^2 \psi$. Upon working out the integral by parts, putting in the appropriate boundary conditions for a bound state function ψ and further taking the 'unperturbed function' ψ to obey the correct Schrödinger equation (3), we obtain, as the reader may verify

$$E_1 = \tfrac{1}{12} \frac{h^2}{\alpha^2} \langle (E - V)^2 \rangle. \tag{13}$$

My argument [3] is that we can produce this shift automatically by using an effective potential term

$$VE = \tfrac{1}{12} \frac{h^2}{\alpha^2} [V(x) - E]^2 \tag{14}$$

to replace *VP*. This has two advantages. First, it allows for *VP* without using an awkward D^4 operator. Second, it requires only a trivial change in the program which executes the method of §2; we simply work out $f = \alpha^{-1}(V - E)$ as before, but use $f + \tfrac{1}{12}h^2 f$ instead of f in the rest of the calculation. This really is a minimal adjustment of the program! Although I discovered (after inventing this method) that a few authors [e.g. 4] had got somewhere within sight of this simple idea in their discussion of errors, nobody seemed to have noticed how to relate it to first-order perturbation theory. Indeed, some numerical analysts still apparently feel that my argument is unsound, even though it is second nature to a physicist brought up on perturbation theory. I have looked at the theory many times, carefully checking how the quantities depend on h. No matter how I slice the cake, I still arrive at my simple result *provided* that the potential $V(x)$ is a smooth non-singular one, so that the boundary terms in the partial integration of the E_1 integral all vanish. The *empirical* test of the method is to see whether using the extra term produces an eigenvalue error which varies as h^4 instead of h^2. It does.

10.4 Some numerical results

Here are some results for the even ground state of the Schrödinger equation

$$-D^2 \psi + x^2 \psi + x^4 \psi = E\psi \tag{15}$$

which was treated by the hypervirial-perturbation approach in §9.7. Method 1 is the simple method (K = 0 in the program). Method 2 is the method of §10.3 (K = $h^2/12$ in the program). The following tables give the Richardson analysis of the results

h	E		
Method 1			
0.08	1.391 320 12	235 171	
0.04	1.392 093 81	235 165	235 165
0.02	1.392 287 19		
Method 2			
0.08	1.392 349 36	5165	
0.04	1.392 351 51	5164	5165
0.02	1.392 351 63		

Method 2 clearly works very well even without the result at h = 0.02, which involves the longest calculation. Nevertheless, method 1 gives a quickly converging table of results for many simple potentials and is quite useful.

10.5 Numerov's method

I now return to the one-dimensional equation (3) and use the short notation $G(x)$ for the quantity $\alpha^{-1}[V(x) - E]$. Since $D^2 \psi = G\psi$ and $D^4 \psi = D^2(G\psi)$ we can use the finite-difference operator δ^2 to replace both second derivatives. The result is that equation (4) *including VP* becomes

$$\delta^2 [\psi - \tfrac{1}{12} h^2 G\psi] = h^2 G\psi + 0(h^4). \tag{16}$$

This equation, with the $0(h^4)$ terms neglected, is the basis of the Numerov method, although in the literature this method is usually employed in conjunction with various fairly complicated matching procedures to calculate eigenvalues. It is easy, however, to use it together with the *EP* calculation and the node counting which I have already described in the preceding sections. When the Numerov equation is used in that way it yields energies which have a leading error term of order h^4, although several authors who have used the Numerov method with varying h have overlooked the easy possibility of improving their energy values by Richardson extrapolation. In the paper [3] in which I invented the method of §10.3 I also pointed out that only a slight modification of the Numerov equation is needed to allow perturbatively for the $O(h^4)$ term. This yields a modified method which gives an energy error of order h^6, but only for smooth non-singular potentials. The Numerov equation will work for potentials such as the Coulomb one, or the centrifugal term $\frac{1}{2}l(l+1)r^{-2}$, which are singular at the origin. Introducing the variables $R(x)$ and $F(x)$ as in previous sections we find the equation

$$F(x)g(x+h) = \frac{F(x-h)}{R(x-h)} g(x-h) + \tfrac{5}{6}G(x) + \frac{G(x+h) + G(x-h)}{12} \quad (17)$$

with

$$g(x) = 1 - \tfrac{1}{12}h^2 G(x). \quad (18)$$

This equation is not much more difficult to use than the simple ones of preceding sections. The main variation needed is the storage of a set of three gs, with the updating transformation

$$g(x+h) \rightarrow g(x) \rightarrow g(x-h) \quad (19)$$

each time x advances by a steplength h. I leave the detail to the reader; the Numerov method is widely discussed in the literature and my main aim has been to describe the slightly unorthodox methods which arise when perturbation theory is used to simplify microcomputer work. Indeed I regard the material of §§9.4, 10.3 and 12.4 as clear examples of what can be accomplished if we take textbook perturbation theory as a tool to be *used* in exploring problems instead of a 'cut and dried' set of standard prescriptions to be routinely applied. Even the simple first-order energy formula has some life left in it, as I hope that I have demonstrated!

10.6   The radial equation

The radial equation can be taken in two forms, as I pointed out in §7.2. (See also [5].) If $R(r)$ is calculated then a centrifugal term must be included in the potential to fix the angular momentum value. In the $\phi(r)$ equation, however, which takes the form

$$-\tfrac{1}{2}rD^2\phi - (l+1)D\phi = (E-V)r\phi \quad (20)$$

(after multiplying by r) there is no centrifugal term included in V. If V is quoted as a polynomial or power series then the power series method can be used to find the eigenvalues. To use the finite-difference method for the equation we can make the lowest order replacements

$$2hD\phi(x) = \phi(x+h) - \phi(x-h) \tag{21}$$

$$h^2 D^2 \phi(x) = \phi(x+h) + \phi(x-h) - 2\phi(x) \tag{22}$$

after which the ϕ equation takes the form

$$[r + (l+1)h]\phi(r+h) + [r - (l+1)h]\phi(r-h)$$
$$= 2r\phi(r) + 2rh^2[V(r) - E]\phi(r). \tag{23}$$

On using the variables $R(r)$ and $F(r)$, the former one not to be confused with the traditional radial wavefunction, we find

$$[r + H]F(r) = [r - H]\frac{F(r-h)}{R(r-h)} + 2[V(r) - E]r \tag{24}$$

where $H = (l+1)h$ is the only quantity which explicitly involves the angular momentum l. Not only is it not necessary to include a centrifugal term in $V(r)$, it is not even necessary to bother very much about starting conditions. By starting at $r = H$ we ensure that the first term on the right vanishes whatever we say about the initial F value, so we could arbitrarily set $R = F = \psi = 1$ at $r = H - h$ without disturbing the rest of the calculation. The rest of the paraphernalia (use of two E values, node counting, calculation of EP) is just as it was for the previous program. Indeed with a little thought the case of an even potential in one dimension can be treated using the radial equation program. By setting $l = 0$ we get the s state solution, which also is appropriate to an odd solution for the one-dimensional problem. By setting $l = -1$, with $F(0) = \frac{1}{2}[V(0) - E]$ and starting at $r = h$ we get results appropriate to even solutions in one dimension (you will have to think about it for a while!).

I invented the method described above so as to simplify the starting procedure and to avoid the use of a centrifugal potential term. The eigenvalues obtained are in error by a leading term of order h^2, requiring a Richardson extrapolation process to convert them to very accurate energies. There does seem to be a price for the simplicity of the method; so far I have not been able to find a simple one-line modification which converts the process to an h^4 one, although I will be happy if any reader can see how to do it! The program is as shown below.

```
10 INPUT L, H : H2 = H * H
20 N = L : N1 = N + L + 1 : N2 = N − L − 1
30 R1 = 1 : R2 = 1 : W1 = 1 : W2 = 1 : S = 1
40 INPUT E, DE, Q
50 N = N + S : N1 = N1 + S : N2 = N2 + S : X = N * H
60 G1 = X/5 − (1/X) − E : G2 = G1 − DE
70 F1 = (G1 * 2 * N + N2 * F1/R1)/N1
80 F2 = (G2 * 2 * N + N2 * F2/R2)/N1
```

```
 90  R1 = 1 + H2 * F1 : R2 = 1 + H2 * F2
100  M = 1 : IF ABS (W2) > 1E30 THEN M = 1 E − 6
110  W1 = W1 * R1 * M : W2 = W2 * R2 * M
120  Q1 = Q1 + (1 − SGN(R1))/2 : Q2 = Q2 + (1 − SGN(R2))/2
130  IF Q2 < Q THEN 50
140  EP = E + DE/(1 − W2/W1)
150  PRINT "∧∧" EP; PRINT Q1, Q2 : GOTO 50
```

The integers N1 and N2 are used to represent $r \pm (l + 1)h$, a factor h being cancelled throughout equation (24). The usual node counter and overflow control are included. The potential (in line 60) has been set as $0.2r - r^{-1}$, a perturbed hydrogen atom potential. In the Richardson table below I show what the program gives for the s state $(l = 0)$ ground state energy. The ratio of the two energy differences from the E column is 3.994, showing the h^2 form of the leading error term.

h	E		
0.02	−0.235 586 135	564 7374	
0.04	−0.235 402 417	564 6985	7400
0.08	−0.234 668 713		

The program given above, although simple and only of h^2 accuracy, is reliable and widely applicable. For example, I have used it in an investigation of singular potentials, with form r^{-3}, r^{-4}, etc at the origin [2]. The resulting smooth Richardson table suggested that earlier work of other authors, who used another computer method, was possibly based on a faulty program. To my relief, this has recently been independently verified by a third party. There are many ways in which the methods described in this chapter can be modified for particular tasks. I try to cover some of them briefly in the exercises.

The structure of the finite-difference program is fairly simple but I give opposite a flowchart to emphasise the main points of the program.

10.7 Further applications

In this chapter I have stuck to one-particle problems, since these can be treated adequately using microcomputers. Several authors [e.g. 6, 7] have discussed how to use finite-difference methods for several-electron atoms, using large computers; it will be interesting to see whether their procedures can be simplified sufficiently to work on microcomputers. In the case of atoms or ions in which one or two

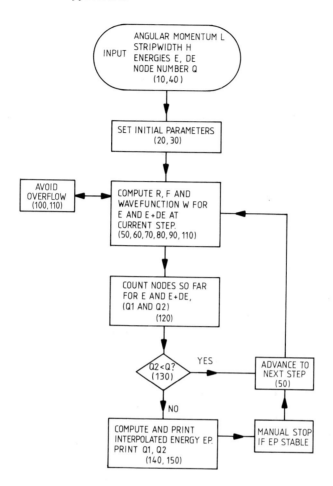

electrons occupy a valence shell outside a filled shell core it is quite common to use an effective single-particle potential to describe the motion of the outer electrons. Many authors [8, 9, 10, 11] have given such effective potentials, and it is perfectly feasible to treat them using the simple methods of this chapter, although most of the authors used a matrix variational method with a few basis functions.

Hajj [12] has recently described an approach to the two-dimensional Schrödinger equation which uses finite-difference methods and banded matrix methods and which may turn out to be of use on small computers. He applies his method to get the s-limit energy of the helium atom (which I discuss in §12.3) and he gets accurate energies by using Richardson extrapolation of results obtained with different stripwidths in the finite element calculation.

Exercises

1. Write down the Schrödinger equation for two different energy values but with the same potential. By doing an appropriate integral between successive nodes for one of the wavefunctions, show that the wavefunction with the greater energy must have the greater number of nodes.

2. How would you use the simple methods of this chapter for:

    (a) A potential with a slope discontinuity at $x = d$?
    (b) A potential such as $x + x^4$ which has bound states but is not of even parity?
    (c) A radial problem with the Dirichlet condition $\psi(d) = 0$?

3. When doing ordinary integrals it is sometimes useful to change variables. Consider the change $r = x^2$ for the ϕ equation, equation (20). How do the equations of the theory and the program change? What advantage is there in making this (or other) changes of variable?

Solutions

1. If we call the function-energy pairs (f, F), (g, G) then we find that the following result holds, if we take the kinetic energy as $-D^2$;

    $$\frac{d}{dx}(fg' - gf') = (F - G)fg. \qquad (25)$$

    Integrating this between two successive nodes of g gives

    $$(G - F) \int_1^2 fg \, dx = f_1 g_1' - f_2 g_2'. \qquad (26)$$

    If we suppose that g is positive between the nodes, then $g_1' > 0$ and $g_2' < 0$. Looking at the signs of the terms in the equation we conclude that if $G < F$ it is impossible for f to have the same sign throughout the region between the nodes of g. There must be one or more nodes of f in the region if $F > G$. Applying this argument to the kind of problems studied in this chapter leads to the conclusion that the number of nodes increases with the energy, and this idea is the basis for the node counting statements which I have included in my programs.

2. (a) The h values used are chosen to make $x = d$ fall on a strip edge; the h^2 methods then work in the same way as for a smooth potential, although the theory behind the h^4 method of §10.3 is no longer valid. (Try it empirically and see what happens!)

(b) If the origin is taken at $x = -D$, with D large, then it will be *almost* exact to set $\psi(-D) = 0$ and proceed forwards from the displaced origin as though treating an odd parity state. Alternatively the radial equation with $l = 0$ can be used. The essential point is to get the potential right at each step, but the only major change in the program is just putting in the coordinate shift correctly to start off the process. For excited states it is easy to increase D slightly to check that this doesn't affect the energy, just in case D was not taken to be large enough to be in the tail of the eigenfunction.

(c) The quantity EP is worked out just *once*, at $r = d$, although this will be officially at $r = d - h$ in the program if it works out $R(d - h)$ first. The value of $R(d - h)$ produces $W(d)$ in my programs.

3. Setting $r = x^2$ gives $dr = 2x\,dx$. Tracing this through the successive differentiations gives the modified ϕ equation

$$xD^2\phi + (2l + 1)D\phi = (V - E)x^3\phi \tag{27}$$

which produces the F equation

$$[x + H]F(x) = [x - H]\frac{F(x - h)}{R(x - h)} + (V - E)x^3 \tag{28}$$

with $H = (l + \frac{1}{2})h$. The modifications to the program are very slight, which is one of the reasons why I chose the form of my programs. At large x, particularly for excited states, the computing time is decreased.

Notes

1. P J Cooney, E P Kanter and Z Vager 1981 *Am. J. Phys.* **49** 76
2. J Killingbeck 1982 *J. Phys. B: At. Mol. Phys.* **15** 829
3. J Killingbeck 1979 *Comput. Phys. Commun.* **18** 211
4. H C Bolton and H I Scoins 1956 *Proc. Camb. Phil. Soc.* **52** 215
5. J Killingbeck 1977 *J. Phys. A: Math. Gen.* **10** L99
6. C Froese 1963 *Can. J. Phys.* **41** 1895
7. N W Winter, A Laferriere and V McKoy 1970 *Phys. Rev.* **2** A 49
8. J D Weeks and S A Rice 1968 *J. Chem. Phys.* **49** 2741
9. G Simons 1971 *J. Chem. Phys.* **55** 756
10. P S Ganas 1980 *Mol. Phys.* **39** 1513
11. T. Seifert *Ann. Phys., Lpz* **37** 368
12. F Y Hajj 1982 *J. Phys. B: At. Mol. Phys.* **15** 683

11 One-dimensional model problems

11.1 Introduction

In chapter 1 I pointed out the value of microcomputers in the teaching of quantum mechanics, where they allow the use of numerical illustrative examples alongside the exposition of the algebraic formalism. In this chapter I give two examples of educational value. They both involve simple one-dimensional calculations which can be treated numerically, but they illustrate basic principles which are of value in quantum chemistry and in solid state physics. The first example illustrates the use of the Born–Oppenheimer approximation in the theory of chemical bond formation, and uses the simple finite-difference methods developed in chapter 10. The second illustrates the formation of energy bands for electrons moving in a periodic potential, and uses the recurrence relations for tridiagonal matrices which were explained in chapter 8.

11.2 A one-dimensional molecule

Suppose that the potential $-(r + a)^{-1}$ with $a > 0$ is used to replace r^{-1} in the hydrogen atom Schrödinger equation. Solving the radial equation for the radial function $R(r)$ (§7.2) will give the usual hydrogen atom energy levels in the limit $a \to 0$ if we use the traditional boundary conditions $R(0) = R(\infty) = 0$. For s states this is the same as looking for odd bound states for the one-dimensional potential $-(|x| + a)^{-1}$, with the boundary conditions $\psi(\pm\infty) = 0$. For the one-dimensional problem, however, we can find even solutions as well, and several authors [1, 2] have discussed the way in which the energy levels behave in the limit $a \to 0$. In particular there is an even ground state with an energy approximately equal to $-2 \ln^2(-4a)$ when a is very small and positive. This ground state energy tends to $-\infty$ as $a \to 0$. However, on setting a exactly equal to zero and

140

solving the problem, we find that the even parity solutions cannot be made smooth at the origin because of the potential singularity, so the odd parity (hydrogenic) levels become the only acceptable 'physical' ones.

In this section I use the potential $-(|x| + a)^{-1}$ not for its intrinsic interest but as a simple potential which illustrates the ideas behind the theory of chemical bond formation. I consider a one-dimensional system which is analogous to the H_2^+ hydrogen molecule ion, but I avoid singularity problems by supposing the interaction potentials between the electron and the two protons to have a non-zero a value, although the full Coulomb ($a = 0$) potential is retained for the inter-proton interaction. I used the Schrödinger equation

$$-\tfrac{1}{2}D^2 \psi + [(2d)^{-1} - V(x + d) - V(x - d)]\psi = E\psi \tag{1}$$

with

$$V(x) = -(|x| + a)^{-1} \tag{2}$$

and the value $a = 1$ to calculate the illustrative numbers quoted below. The one-dimensional molecule has two fixed protons with separation $2d$ and one electron which is attracted to both protons by the modified potential which I use. Taking the origin halfway between the protons gives a potential of even parity, so we can look for even or odd solutions using the simple programs of §10.2. Strictly speaking, it is against the idea of the uncertainty principle to give exact values to the proton positions and neglect their momenta as well! However, the procedure commonly used is to do this first and then later let the nuclei move by solving another (nuclear) Schrödinger equation in which the nuclear potential function is the energy curve obtained from the electronic calculation. This procedure is often called the Born–Oppenheimer approximation and is a standard one, although in recent years very small correction terms to the resulting energies have been looked at.

I found the lowest even and odd energy levels of the Schrödinger equation (1) using the simple methods of §10.2 and the results are given below for a small range of d values around the one at which the even state has an energy minimum. The value for $d = \infty$ is just the ground state even parity energy for the one-centre potential $-(|x| + 1)^{-1}$.

d	$E(\text{even})$	$E(\text{odd})$
1.6	−0.509 997	−0.3196
1.7	−0.511 780	−0.3413
1.8	−0.512 492	−0.3606
1.9	−0.512 423	−0.3769
2.0	−0.511 796	−0.3914
2.1	−0.510 780	−0.4037
∞	−0.5	−0.5

The energy minimum, $-0.512\,546$ at $d = 1.84$, can be found by Newton–Gregory interpolation (§4.2) or by a more detailed computation. The Born–Oppenheimer approximation would next take the numbers in the E column as giving a potential energy function and would use that function in a Schrödinger equation describing the motion of the nuclei (for which a mass would have to be specified). The result would be oscillatory motions (with slight anharmonicity) around the potential minimum. The distance $2d = 3.68$ is, of course, the chemical bond length for my hypothetical molecular ion. In the LCAO (linear combination of atomic orbitals) approach the ground state wavefunctions ψ_1 and ψ_2 for the electron ground state on nucleus 1 and 2, respectively, would be used. The linear combinations $\psi_1 \pm \psi_2$ would be formed, to give even and odd parity molecular orbitals respectively. Working out the energy expectation values for these two functions would produce energy versus d curves which are qualitatively the same as those from the numerical integration, although not as accurate because of the approximate form of the trial wavefunctions. By using more flexible trial wavefunctions, of course, the variational approach can be made quite accurate; it will always give upper bounds to the eigenvalues, since the two states being studied are ground states, one for even parity and the other for odd parity. The results show that the odd (anti-bonding) state does not give an energy minimum in the d range studied; in fact, it does not give one at any d value (except infinity). If the molecular ion were to undergo an electronic transition from the ground state to the odd state then it would dissociate spontaneously to poduce a neutral atom (proton plus electron) and a proton.

11.3   A one-dimensional band problem

As an example of a Schrödinger equation involving a periodic potential I take the equation

$$-D^2 \psi + \lambda \cos x\psi = E\psi. \tag{3}$$

The potential is a very simple one, corresponding to a lattice with unit cell length 2π. For a free particle, with $\lambda = 0$, appropriate travelling wavefunctions would be of form $\exp(ikx)$, with energy k^2. The functions $\cos kx$ and $\sin kx$ would also be eigenfunctions with energy k^2. When λ is non-zero the perturbation $\cos x = \frac{1}{2}[\exp(ix) + \exp(-ix)]$ will link together waves which have k values differing by ± 1. The result of starting with one wave and then combining it with all waves which can be coupled to it (and to one another) is to produce a function of form

$$\psi = \exp(ikx) \sum_{-\infty}^{\infty} A_N \exp(iNx). \tag{4}$$

This is the product of exp (ikx) and a function which is periodic in the lattice (here of cell length 2π). The result that the eigenfunctions take this product form is usually called Bloch's theorem.

If the eigenfunction ψ is postulated to take the form (4) then by inserting it in the Schrödinger equation we quickly obtain the recurrence relation

$$[(N + k)^2 - E]A_N + \tfrac{1}{2}\lambda[A_{N+1} + A_{N-1}] = 0. \tag{5}$$

This leads to a tridiagonal matrix problem (§8.3) with the novelty that it is infinite in both directions, since N ranges from $+\infty$ to $-\infty$. One way to handle this is to take a $(2M + 1) \times (2M + 1)$ matrix ranging from $N = M$ to $N = -M$ and find its eigenvalues (for given k) by the usual methods for tridiagonal matrices (§8.3). As M is increased the lower eigenvalues will approach their limiting values. For special cases it is possible to make the matrix infinite in one direction only, which further simplifies the computations; the clue is to notice that a product of form $\cos nx \cos x$ involves only cosines, while $\sin nx \cos x$ involves only sines. The reader may check that the following families of terms are closed in that the Hamiltonian does not link them to functions outside the family:

$$\{\cos Nx\}, \{\sin Nx\}, \{\cos (N + \tfrac{1}{2})x\}, \{\sin (N + \tfrac{1}{2})x\} \tag{6}$$

(In all cases N goes from 0 to $+\infty$).

The first two families correspond to Bloch functions with $k = 0$, the last two to Bloch functions with $k = \tfrac{1}{2}$. The value $k = \tfrac{1}{2}$ is at the boundary of the first Brillouin zone. This zone stretches from $k = \tfrac{1}{2}$ to $k = -\tfrac{1}{2}$ and every Bloch state can be assigned a k value in this zone. For example, an unperturbed wave with $k = \tfrac{3}{4}$ is coupled by the potential to the $k = -\tfrac{1}{4}$ wave and so can be formally classified as giving an excited state with $k = -\tfrac{1}{4}$. By putting a sum of $\cos (N + \tfrac{1}{2})x$ terms in the Schrödinger equation we obtain a tridiagonal matrix eigenvalue problem with the associated determinant

$$\begin{vmatrix} (\tfrac{1}{4} + \tfrac{1}{2}\lambda - E) & \tfrac{1}{2}\lambda & 0 & \cdot \\ \tfrac{1}{2}\lambda & (\tfrac{9}{4} - E) & \tfrac{1}{2}\lambda & \cdot \\ 0 & \tfrac{1}{2}\lambda & (\tfrac{25}{4} - E) & \cdot \\ \cdot & \cdot & \cdot & \cdot \end{vmatrix} \tag{7}$$

Using the $\sin (N + \tfrac{1}{2})x$ family yields a determinant which differs only by having $-\tfrac{1}{2}\lambda$ instead of $\tfrac{1}{2}\lambda$ in the first element. For small λ this means that there is a band gap of width λ in the energy level diagram at $k = \tfrac{1}{2}$. Since the energy $E(k)$ is monotonic as a function of k (in each band) it follows that there are no energy levels of any k in the gap region. The determinant above is very easy to deal with by the recurrence method of §8.3. We simply set $D(0) = 1, D(1) = (\tfrac{1}{4} \pm \tfrac{1}{2}\lambda - E)$

and use the relation

$$4D(N) = [(2N-1)^2 - 4E]D(N-1) - \lambda^2 D(N-2) \tag{8}$$

to calculate $D(10)$, for example, for various trial energies until the E values which render it zero are located. I give below a program for the ZX-81 which will take an initial energy estimate and iterate automatically to produce a root. The symbol L is used for λ and the value of K is input as 1 for the cos states and as -1 for the sin states. For small λ the location of the roots is easy to guess fairly closely by looking at the determinant, so the iteration converges quickly. The program is as follows (it looks 'wrong', but it seems to work better with F as official input; try it and see).

```
10 DIM D(10)
20 INPUT L
30 INPUT K
40 LET D = 0
45 LET E = 0
50 INPUT F
60 LET D(2) = 0.25 + K * L/2 − E
65 LET D(1) = 1
70 FOR N = 3 TO 10
80 LET M = (2 * N − 3)
85 LET M = M * M − 4 * E
90 LET D(N) = (M * D(N − 1) − L * L * D(N − 2))/4
100 NEXT N
105 LET P = (D(10) * F − D * E)/(D(10) − D)
115 LET F = E
120 LET D = D(10)
130 PRINT P
135 LET E = P
140 GOTO 60
```

At $\lambda = 0.2$ I found the {cos, sin} energy pairs {0.344 75, 0.145 25} and {2.252 75, 2.252 25}. These results show the energy gap of roughly λ for the lowest pair and also show a very much smaller but non-zero splitting for the second pair of levels with $k = \frac{1}{2}$. If the potential is a sum of terms of form $V(n) \cos nx$ then this second pair will have an energy gap roughly equal to $V(2)$. In general each $V(n)$ gives a direct splitting of size $V(n)$ for the nth level pair at $k = \frac{1}{2}$ plus a weak higher order effect on the splittings of the other pairs at $k = \frac{1}{2}$. In a three-dimensional problem the quantity k becomes a wave vector $\mathbf{k}$ and the Brillouin zone becomes a parallelopiped around the origin in k-space. The coefficients A involve three integers and so become $A(N_1, N_2, N_3)$, while the component terms of the potential similarly have three labels. The resulting energies can be found by setting up the matrix eigenvalue problem using a plane wave set exp ($\mathrm{i}kr$) as basis. As our example shows, the calculation is more like a Hill determinant one than an inner product matrix calculation (§8.3). The various Fourier components in the potential produce energy splittings for the states belonging to **k** vectors on the Brillouin zone boundaries. The theory sketched here is usually called the nearly-free electron (NFE) theory. Since it takes the lattice potential to be a weak perturbation, it was supposed for a long time that it would not be of much use for electrons in a real metal, where the periodic potential is estimated to be strong. However, modern work has shown that the approach can be used for the conduction electrons in many metals if a weak effective potential (the so-called pseudopotential) is used in the one-electron Schrödinger equation. It turns out that the basic requirement that the conduction electron wavefunctions must be orthogonal to the core electron wavefunctions of the inner shells acts as a constraint in the calculation of the conduction electron wavefunctions. This constraint has the effect of cancelling much of the strong periodic potential and making the conduction electron wavefunctions behave as if they were controlled by a weak pseudo-potential. I refer the reader to solid state theory texts for the details; see e.g. [3]. In practice the procedure used is often semi-empirical; workers simply find a weak potential which fits their experimental results when used in the NFE formalism, leaving the *a priori* justification to more scrupulous logicians. The same applies to crystal field theory and spin Hamiltonian theory. Indeed, the same fate of widespread and uncritical application befalls many of the ideas of theoretical physics. (I make no complaints about this, of course: I am partly able to write this book because of the widespread and uncritical application of large computers to problems where microcomputers would suffice!)

Notes

1. F Gesztesy 1980 *J. Phys. A: Math. Gen.* **13** 867
2. L K Haines and D H Roberts 1969 *Am. J. Phys.* **37** 1145
3. W A Harrison 1970 *Solid State Theory* (New York: McGraw-Hill)

12 Some case studies

12.1 Introduction

In chapter 1 I noted that this book is filled with case studies in which a micro-computer method evolves out of an integrated theoretical study of some problem of classical or quantum mechanics. In this final chapter I have gathered together several detailed case studies which have some relevance to problems still being treated in the research literature. The main criterion which I used to pick a topic for this chapter was that it should illustrate the blending of several different theoretical and calculational methods in one problem. This ensures that the topics will show the value of combining analysis with computation; I take it that this direct demonstration will be much more effective than any general argument on my part. The helium atom problem of §§12.2 and 12.3 involves the analytic evaluation of Coulomb integrals, the use of the variational principle and the method of Monte-Carlo optimisation. The charmonium problem (§12.4) involves change of scale (§2.5), the series method (§7.4) and perturbation and hypervirial techniques (chapter 9). The methods of chapter 7 and 9, plus some angular momentum theory, are needed for the quadratic Zeeman problem of §12.5. The quasi-bound state calculation of §12.6 uses a modification of the finite-difference methods of chapter 10 together with the Newton–Gregory formula of chapter 4. With these detailed case studies I conclude my submission in defence of the use of microcomputers in the study of quantum mechanics. I hope the reader will by now have begun to agree that the case is a pretty strong one!

12.2 A simple helium atom calculation

The Schrödinger equation (in atomic units) for the helium atom takes the form

$$[-\tfrac{1}{2}(\nabla_1^2 + \nabla_2^2) - 2(r_1^{-1} + r_2^{-1}) + r_{12}^{-1}]\psi = E\psi \tag{1}$$

if we take the nuclear mass to be infinite. If the inter-electron repulsion term r_{12}^{-1} is neglected then the ground state is the 1 s^2 function

$$N \exp\left(-2r_1 - 2r_2\right) \tag{2}$$

where N is a normalisation constant. The spin factor $[\alpha(1)\beta(2) - \beta(1)\alpha(2)]$ should also be included in the wavefunction. Since this singlet spin factor is antisymmetric for the permutation $1 \leftrightarrow 2$, the total wavefunction will be anti-symmetric if the orbital factor is symmetric. For example, the function

$$\psi = \left[\exp\left(-Ar_1 - Br_2\right) + \exp\left(-Br_1 - Ar_2\right)\right] \tag{3}$$

is symmetric, but by varying A and B we can make the expectation value of the energy lower than we could by keeping A and B equal. The energy operator is given in equation (1). The integrals for the expectation values of the various operators are rather tedious to evaluate. I have sketched some of the relevant mathematics in Appendix 1, and merely quote the results here. I denote the kinetic energy operator by T, the nuclear attraction operator by U and the identity operator by $\mathbf{1}$. The various expectation values are then as follows.

$$\langle \mathbf{1} \rangle = \tfrac{1}{8}(AB)^{-3} + 8(A + B)^{-6} \tag{4}$$

$$\langle U \rangle = -\tfrac{1}{4}(A^{-2}B^{-3} + B^{-2}A^{-3}) - 16(A + B)^{-5} \tag{5}$$

$$\langle T \rangle = \tfrac{1}{16}(A^2 + B^2)(AB)^{-3} + 8AB(A + B)^{-6} \tag{6}$$

$$\langle r_{12}^{-1} \rangle = 2\left[I(2A, 2B) + I(2B, 2A) + 2I(A + B, A + B)\right]. \tag{7}$$

Here the integral

$$I(a, b) = (2b + 8a)a^{-2}(a + b)^{-4} \tag{8}$$

is of the $I(1, 2; a, b)$ type mentioned in Appendix 1. According to the variational principle what we have to do next is to vary A and B in order to minimise the energy expectation value. This means that we have to minimise the ratio $\langle T + U + r_{12}^{-1} \rangle / \langle \mathbf{1} \rangle$ formed from the quantities which I have quoted above. I have already gone a little further than most textbooks by treating the case $A \neq B$; the usual procedure is to set $A = B$ and show that $A = 1.6875$ gives a minimum energy of -2.8477, whereas $A = B = 2$ gives $E = -2.7500$. This latter result is, of course, just the energy expectation value for the unperturbed ground state function. I now want to go a little further still, since I am writing about micro-computers; I shall attack the problem by using an interesting method which is well suited to microcomputers with random number facilities.

12.3 Monte-Carlo optimisation

While the idea behind Monte-Carlo optimisation is not new, the recent work of

Conley [1] has shown how it can be used in a variety of problems. I recommend his book to my readers; it gives many BASIC programs which will run on small machines, although I suspect that he used them on larger machines than those about which I am writing here. Conley does not deal with quantum mechanical problems, but I intend to use the Monte-Carlo optimisation approach to deal with the helium atom problem which I set up in the preceding section. The idea is simple. If we suppose that the optimum A and B values will probably be in the region 1 to 3, since the unperturbed function has $A = B = 2$, then we might try varying A and B over the whole region to get the minimum. If we search with increments of 0.01 in A and B it may take a long time to find the minimum if we make an unlucky choice of the starting point. The Monte-Carlo optimisation process simply spatters the sample points randomly but uniformly over the whole permitted 'square' of (A, B) pairs, and quickly gives an approximate location for the minimum. The search square can then be made smaller to get a more precise location of the minimum, and so on. Most microcomputers, when they produce a random number by a RND(X) function, give a number between 0 and 1, so to produce numbers between 1 and 3 we have to use a quantity $1 + 2 * RND(X)$, i.e. we have to shift and expand the range of the random numbers. Here is the program which I used to minimise the energy. I have not given the specific lines in which the function of A and B is evaluated, because any function of A and B could be inserted in that part of the program.

```
10 INPUT A0, A1, B0, B1
15 Q = 1 : P = 100
20 A = A0 + A1 * RND(Q)
30 B = B0 + B1 * RND(Q)
   [Calculate E(A, B)]
80 IF P < E THEN 100
90 P = E : PRINT E, A, B
100 GOTO 20
```

On a machine without a random number generator the following modified instructions could be used to produce the random number sequence

```
15 Q = 1 : P = 100 : R = 0.5
20 GOSUB 200 : A = A0 + A1 * R
30 GOSUB 200 : B = B0 + B1 * R
200 R = (π + R) ↑ 5 : R = R − INT(R)
210 RETURN
```

There are many ways of producing a sequence of quasi-random numbers in line 200; this one was quoted by G Relf [2] in an article on maze games. The parameters A0, A1 are chosen so that the A value is spread uniformly over the region A0 to A0 + A1, and similarly for B. The 'lowest value so far' is printed out, with its (A, B) value, in line 90. Because of the randomness involved not every run gives identical results, but here are some typical results. The input parameters are shown on the left and the output values after a minute or so are shown on the right.

1, 2, 1, 2	−2.8725	2.253, 1.231
1, 0.5, 2, 0.5	−2.8746	2.144, 1.244
2, 0.1, 1.1, 0.2	−2.8756	2.169, 1.196
2.1, 0.1, 1.15, 0.1	−2.8757	2.179, 1.189

The energy minimum is −2.8757. By setting B = A in the program I obtained the usual textbook minimum of −2.8477 at A ≅ 1.7.

The Hartree-Fock approach for this problem tries to find the orbital function ϕ such that the product function $\phi(r_1)\phi(r_2)$ gives the minimum energy. The best Hartree-Fock energy, −2.8617, is not as good as that obtained here, although it is necessarily better than the result for the $A = B$ constrained calculation. One 'clever' idea would be to try a function based on (3), but with Hartree-Fock functions instead of exponentials, so that the function takes the form $[\phi(Ar_1)\phi(Br_2) + \phi(Ar_2)\phi(Br_1)]$. It turns out that this gives a worse result than that which I have just obtained using exponential functions! This at first sight unexpected result can be explained in terms of the mathematical properties of Hartree-Fock functions [3]; this was a case in which unusual numerical results suggested new theoretical investigations. I have also applied the (A, B) parameter approach to a $2p^2$ configuration, with ^3P, ^1D and ^1S terms, and other authors have used it for atoms with more than two electrons [4]. In terms of a configuration interaction approach the simple function (3) roughly allows for the coupling of the unperturbed $1s^2$ function to excited state functions involving s orbitals. Very detailed calculations suggest that even with a complete set of s orbital functions the lowest energy attainable (the so-called s limit) is −2.8790, so the two parameter calculation is really quite efficient. The use of s functions only treats radial correlation, the tendency of the electrons to avoid one another by having different r values. Angular correlation, the tendency of the electrons to keep away from one another in angle coordinates, can only be described by using p, d and higher basis functions in a variational calculation. With these included the energy can be brought down to around −2.9032. Some simple ways of using the Hylleraas principle (§9.3) to estimate the energy lowering caused by including p, d and higher basis functions are described in my early papers [5, 6]. These works provide good examples for microcomputer calculations, although I actually used simple electronic calculators for the original work.

12.4   The charmonium problem

In recent years an attempt has been made to explain the failure to detect free
quarks by supposing that quarks and antiquarks are bound in pairs. The bound
quark-antiquark pair (the so-called charmonium system) is supposed in the
simple models to be described by a Schrödinger equation of form

$$-\alpha\nabla^2\psi - \frac{\beta}{r}\psi + \mu r^M\psi = E\psi \tag{9}$$

where M is a positive integer and α involves the reduced mass of the particle pair.
r will be the interparticle distance, with the centre of mass motion having been
separated out, but we can proceed mathematically as though we were treating
a single-particle Schrödinger equation. The confining potential term μr^M (with
$\mu > 0$) will ensure that all the states are bound states. The first thing to do is to
use the coordinate scaling idea explained in §2.5 and so avoid having to use
three parameters α, β and μ. By making a change of length scale $r \to Kr$ we find
that the following relationship holds,

$$E(\tfrac{1}{2}, 1, \lambda) = K^2 E(\alpha, \beta, \mu) \tag{10}$$

with

$$\beta K = 2\alpha : \lambda = \mu K^{M+2}. \tag{11}$$

This result means that all we need to do is to compute the energies and other
properties for the case $\alpha = \tfrac{1}{2}$, $\beta = 1$ and vary only λ. By inserting a factor
$Y_l^m(\theta, \phi)$ or Y_l in the wavefunction (§7.2) we arrive at a radial equation in the
R or ϕ form. If in the ϕ equation

$$-\tfrac{1}{2}D^2\phi - (l+1)r^{-1}D\phi + V\phi = E\phi \tag{12}$$

we insert the postulated form

$$\exp(-\gamma r)\ \Sigma\ A(n)r^n \tag{13}$$

then we obtain the recurrence relation

$$\tfrac{1}{2}(N+2)(N+3+2l)A(N+2) = \gamma(N+2+l)A(N+1)$$
$$-(E+\tfrac{1}{2}\gamma^2)A(N) - A(N+1) + \lambda A(N-M). \tag{14}$$

It should be clear how to modify this equation when the potential is a sum of
various powers. I shall present a program which applies the power series approach
for the potential $-r^{-1} + Ur + Vr^2$, so that the term $\lambda A(N-M)$ above is replaced
by $UA(N-1) + VA(N-2)$. The idea is the same as for the oscillator problems
of §7.4; $\phi(r)$ is worked out at large r for two trial energies and an interpolation
gives the energy which would make $\phi(r)$ zero. γ can be varied to speed up

convergence. The various tricks (projected energy calculation, overflow control, etc) are by now familiar to the reader. To cut down on overflow problems the quantities $A(n)r^n$ are denoted by $A(n)$ or $B(n)$ (one for each energy) in the program. (Otherwise the factor r^n on its own can cause an overflow halt when r is large.) The program is written out below, in Pet style, with the $A(n)$ and $B(n)$ declared as arrays, and with the (U, V) pair arbitrarily set at $(0.1, 0)$ in line 25. For these values the ground state energy with $l = 0$, $\gamma = 1$, $r = 10$ is $-0.360\,900\,076$. The reader may use this to check that the program is working property; only about 25 terms of the series are needed to get convergence.

```
10  DIM A(5), B(5) : A(1) = 1: B(1) = 1: N = −2
20  INPUT L, G, X : R = 1: K = 3 + 2 * L
25  U = 0.1: V = 0 : U = U * X ↑ 3 : V = V * X ↑ 4
30  B = (G * L − 1) * X : WA = 1: WB = 1
40  INPUT E, DE : CA = (E + G * G/2) * X * X :
            CB = CA + DE * X * X
45  G = G * X
50  N = N + 1: B = B + G
60  D = (N + 2) * (N + K)/2 : M = 1
70  A(0) = (B * A(1) − CA * A(2) + U * A(3) + V * A(4))/D
80  B(0) = (B * B(1) − CB * B(2) + U * B(3) + V * B(4))/D
90  WA = WA + A(0) : WB = WB + B(0)
100 IF ABS(WB) > 1E30 THEN M = 1E − 6
110 WA = WA * M : WB = WB * M
120 FOR I = 4 TO 1 STEP −1
130 A(I) = A(I − 1) : B(I) = B(I − 1) : NEXT I
140 EP = E + DE/(1 − WB/WA + R) : R = 0
150 PRINT "∧" EP, N : GOTO 50
```

In my original work [7] on the charmonium potential I used the finite-difference method (§10.2) to get the energy values, but have decided to vary the story a little for this book! As I have pointed out in §9.4, the problem of finding $\langle r \rangle$ or $\langle r^2 \rangle$ can be solved by varying U or V slightly and looking at the resulting energy change. States with $l = 0$ have a non-zero wavefunction at the origin. As explained in §9.5 the value of the wavefunction at $r = 0$ can be related to the expectation value of the derivative of the potential. For a potential $-r^{-1} + Ur + Vr^2$ the derivative is $r^{-2} + U + 2Vr$. The value of $\langle r \rangle$ can be found by varying the value of U in the energy calculation, since $\langle r \rangle = \partial E/\partial U$. How can we get $\langle r^{-2} \rangle$? In the traditional radial equation the r^{-2} term is the centrifugal term $\frac{1}{2}l(l + 1)r^{-2}$. For an s state, with $l = 0$, adding a small term λr^{-2} to the potential looks like giving the angular momentum the value 2λ, if λ is very small compared to l. By setting $l = 2\lambda$ in the program above we effectively add

λr^{-2} to the potential and can get $\langle r^{-2} \rangle$ by energy differencing. The trick works mathematically, even though the non-integer angular momentum values used are not 'physical' ones for any allowed quantum mechanical eigenfunctions – a good example of using theory to help simplify computation! For the test case $l = 0$, $\gamma = 1$, $r = 10$ mentioned above the following energies result as l is varied.

l	E
0.001	$-0.359\ 698\ 808$
0	$-0.360\ 900\ 076$
-0.001	$-0.362\ 104\ 289$

These results imply that $\langle r^{-2} \rangle$ is 2.4055, taking only five figures. The unperturbed hydrogen atom, with $U = V = 0$, has $\langle r^{-2} \rangle = 2$, and the result shows how the confining potential $0.1r$ squeezes in the wavefunction so that it piles up more around the origin. With $U = 0.1$ the result for $\langle r^{-2} \rangle + U$ is 2.5055, showing that the squared wavefunction at the origin has multiplied by a factor 1.2528 as U has increased from 0 to 0.1. If the coefficient U is varied (with l and V zero, of course) then the following results are obtained:

U	E
0.101	$-0.359\ 592\ 548$
0.100	$-0.360\ 900\ 076$
0.099	$-0.362\ 208\ 911$

These results give $\langle r \rangle = 1.308\ 18$ to six figures, whereas the unperturbed hydrogen atom ground state has $\langle r \rangle = 1.5$. The results for $\langle r^{-2} \rangle$ and $\langle r \rangle$ show, incidentally, that the distortion of the wavefunction on going from $U = 0$ to $U = 0.1$ cannot be described by a simple scaling-in of the wavefunction. The scaling factor would have to be $1.5/1.308 = 1.147$ to explain the $\langle r \rangle$ ratio and $(2.406/2)^{1/2} = 1.097$ to explain the $\langle r^{-2} \rangle$ ratio. That the scaling-in effect should be greater at large r is physically reasonable, since the perturbing potential $0.1r$ increases with r.

For the Schrödinger equation

$$-\tfrac{1}{2}\nabla^2 \psi - r^{-1}\psi + \lambda r^M \psi = E\psi \tag{15}$$

the hypervirial relations (§9.6) give the following relation between various expectation values for states with angular momentum l:

$$2(N+1)E\langle r^N \rangle + (2N+1)\langle r^{N-1} \rangle$$
$$+ \tfrac{1}{4}N[(N^2 - 1) - 4l(l+1)]\langle r^{N-2} \rangle$$
$$= \lambda(M + 2N + 2)\langle r^{M+N} \rangle. \tag{16}$$

The term involving $l(l+1)$ arises by formally including a centrifugal term $\frac{1}{2}l(l+1)r^{-2}$ in the potential function. The numerical results which I have been quoting above were for the case $M=1$, $\lambda=0.1$. I show below the hypervirial relations for the first few N and for $M=1, l=0$.

$$2E + \langle r^{-1} \rangle = 3\lambda \langle r \rangle$$

$$4E\langle r \rangle + 3 = 5\lambda \langle r^2 \rangle \tag{17}$$

$$6E\langle r^2 \rangle + 5 \langle r \rangle + 3/2 = 7\lambda \langle r^3 \rangle.$$

These results show that from the values of E and $\langle r \rangle$ it is possible to obtain $\langle r^{-1} \rangle$, $\langle r^2 \rangle$, $\langle r^3 \rangle$ and so on. For the case $M=2$ the necessary quantities to 'start the chain' for s states are E, $\langle r \rangle$ and $\langle r^2 \rangle$, and all of these are obtainable using my simple program for the power series method.

It is also possible to apply the hypervirial-perturbation method ($\S9.7$) to the Schrödinger equation (15) and to form Padé approximants to the resulting divergent series by using Wynn's algorithm ($\S6.4$). Detailed tables of numerical results for this approach are given in [7], but, as I have commented previously, the power series or finite-difference methods are more efficient when λ is large. When λ is large it is possible to treat the Coulomb $-r^{-1}$ term as a perturbation, with λr as the unperturbed potential. By starting from the Schrödinger equation

$$-\alpha D^2 \psi + \lambda r \psi = E(\lambda)\psi \tag{18}$$

and making the coordinate change $r = kx$ (as in $\S2.5$) or, more simply, by using a dimensional argument with $\alpha \equiv [E][L]^2$, $\lambda \equiv [E][L]^{-1}$, we find that the energy eigenvalues vary as $(\alpha\lambda^2)^{1/3}$ and that $\langle r^{-1} \rangle$ varies as $(\lambda\alpha^{-1})^{1/3}$ (Try it: these are the combinations with dimensions $[E]$ for energy, and $[L]^{-1}$ length^{-1}.) Now, if I may reintroduce my friend and standby the first-order energy formula of $\S9.2$, the effect of the perturbation μr^{-1} in the potential of equation (18) is to give an energy shift $\mu\langle r^{-1} \rangle$. The energy expression thus must take the form (with α fixed)

$$E = E_0(1)\lambda^{2/3} + \mu E_1(1)\lambda^{1/3} + \ldots \tag{19}$$

where the coefficients $E_0(1)$ and $E_1(1)$ can be found numerically by calculating energies near $\lambda = 1$, $\mu = 0$, using the potential $\mu r^{-1} + r$. Taking the ratio of the first two terms suggests that the 'effective perturbation parameter' in the E series is $\mu\lambda^{-1/3}$, so that the series takes the following form (with $\beta = \mu\lambda^{-1/3}$)

$$E(\lambda, \mu) = \lambda^{2/3} \Sigma E_n(1)\beta^n. \tag{20}$$

The coefficients can be found by calculations in the neighbourhood of $\lambda = 1$: the second-order contribution has a factor λ^0 and so is λ-independent. Suppose now that we start from the traditional energy perturbation series in powers of

λ for the energy $E(\frac{1}{2}, 1, \lambda)$, with $M = 1$. Any Padé approximant of finite order will be a ratio of two polynomials and for $\lambda \to \infty$ it must behave as some integer power of λ, so that it cannot fit to the $\lambda^{2/3}$ behaviour which is required. We thus have a difficulty of principle about using the usual approximants for large λ, although they work quite well numerically when λ is small. It is difficult to find an approximant which works equally well at large and small λ. I should note in passing that the Schrödinger wavefunctions for the case of the potential λr can be expressed in terms of Airy functions. For my money that is not much help, just as it is not much help to know that some other problems can be solved using Bessel functions. When it comes to using numbers we still have to do computations, since the traditional 'named' functions are usually calculable only by power series or integral representations. My standard query in such cases is a simple one; never mind what it is called, what does it *do*? (It probably arises from my fascination with logic and philosophy, in which the name of a thing and the thing itself are carefully distinguished; only mock-scientists think that naming a problem is equivalent to solving it.)

12.5   The quadratic Zeeman effect

When a magnetic field is applied to a hydrogen atom the relevant Schrödinger equation for the orbital motion is of form

$$[-\tfrac{1}{2}\nabla^2 - r^{-1} + \tfrac{1}{2}\gamma l_z + \tfrac{1}{8}\gamma^2(x^2 + y^2)]\psi = E\psi. \tag{21}$$

Here I take the simplified case of infinite nuclear mass and use atomic units in which the 1s ground state has energy $-\tfrac{1}{2}$ at $\gamma = 0$. The field strength parameter γ is such that $\gamma = 1$ corresponds to a magnetic field of 2.3505×10^4 tesla in experimentalist's units. The extra contribution from the electron spin Zeeman energy can simply be added on when the orbital calculation has been completed. For an s state ($l = 0$) the γ^2 term, the quadratic Zeeman term, is the only one involving the field. It is often neglected in textbook calculations, but has been studied in the recent research literature, since it is thought that there are some situations in which the γ^2 term is not negligible. For example, on some stars enormous magnetic fields are present, and in some solid state materials the electrons have very small effective mass, which is equivalent to having a stronger effective magnetic field.

Even a value of $\gamma = 0.1$ corresponds to a really enormous field by laboratory standards, but leads to only a small mathematical perturbation for the low-lying states of the hydrogen atom. The use of some form of perturbation theory seems appropriate, but the complication is that the perturbing operator has a factor $(x^2 + y^2)$ which is not spherically symmetric. At $\gamma = 0$ each hydrogenic state can be assigned definite l and m quantum numbers. The perturbing term will mix states of different l, but will not change the parity or the m value, since it is of even parity and is rotationally invariant about the z-axis. Since the m value is conserved by the perturbation the term $\frac{1}{2}\gamma l_z$ in the Zeeman operator simply gives a fixed contribution $\frac{1}{2}\gamma m$ to the energy of a state. The technical problem is really to find the energy shift caused by the quadratic term, so I shall refer to that term as 'the perturbation V' in the rest of the discussion. If V is simply inserted into the perturbation formalism of §9.2 then it is possible, but very tedious, to calculate the perturbed functions ψ_n and the energy coefficients. I have done the calculation for the 1s state to get the series

$$E = -\tfrac{1}{2} + 2\lambda - \tfrac{53}{3}\lambda^2 + \tfrac{5581}{9}\lambda^3 \ldots \tag{22}$$

with $\lambda = \frac{1}{8}\gamma^2$ being the coefficient of the term $(x^2 + y^2)$ in V. The series is divergent but can be converted to Padé approximants in λ. I found that the sum $\frac{5}{8}[2/1] + \frac{3}{8}[1/1]$ gives a fair fit to the accurate E values found by other techniques over the range $0 < \gamma < 1$. For other states a similar calculation would be very complicated and I don't think that anyone has tabulated the results systematically.

My own approach [8] is to split V up into two parts, V_1 and V_2. V_1 and V_2 are state-dependent: V_1 is a multiple of r^2, chosen so that $\langle V_1 \rangle = \langle V \rangle$ for the unperturbed state. V_2 has $\langle V_2 \rangle = 0$ for the unperturbed state. (Note how that good old first-order energy formula is coming in again!) I will explain how it works for the case of the 2p$_0$ state ($l = 1$, $m = 0$). This state is really a ground state of the family of states with odd parity and $m = 0$, even though it looks like an excited state in the list of hydrogen atom states; each symmetry species can be treated separately, since the perturbation does not link them. The 2p$_0$ state has the form $z \exp(-r/2)$, apart from a normalisation factor; working out the various expectation values gives the decomposition

$$V(2p_0) = V_1 + V_2$$
$$= \tfrac{1}{20}\gamma^2 r^2 + \tfrac{1}{40}\gamma^2 [3x^2 + 3y^2 - 2z^2]. \tag{23}$$

Now, the term V_2 gives no first-order shift for any state with $l = 1$, $m = 0$; this is a property of the angular integrals involved, or, if you like, a result of angular momentum theory. If we solve the Schrödinger equation for the potential $-r^{-1} + V_1$ we have a radial problem. This yields a perturbed state which still has $l = 1$, $m = 0$, because of the spherical symmetry; we thus allow entirely for V_1, and adding V_2 produces no further energy shift in first order. The point is that we can simply take over the program of §12.4 to deal with the potential $-r^{-1} + \frac{1}{20}\gamma^2 r^2$. The residual effect of V_2 vanishes in first order, while in second order it will slightly depress the energy, since E_2 is always negative for a ground state.

E_2 can actually be closely estimated from the Hylleraas principle of §9.3 as I now explain. By using the series method of §12.3 we can find E, $\langle r \rangle$ and $\langle r^2 \rangle$ and can get higher $\langle r^n \rangle$ by using the hypervirial relations. To find the E_2 value produced by V_2 we start from the eigenfunction of the perturbed radial problem as the 'unperturbed' function. (It does not matter that we could not produce it on request; all we use are the $\langle r^n \rangle$ values.) Calling this function ϕ_0 and the trial function ψ we need to work out the Hylleraas functional (§9.3)

$$2\langle \psi | V_2 | \phi_0 \rangle - \langle \psi | (H_0 - E_0) | \psi \rangle. \tag{24}$$

H_0 and E_0 both refer, of course, to the situation with V_1 included in the Schrödinger equation. If we now take the trial function in the form $f\phi_0$, with f some function of the coordinates, then we quickly see that

$$\langle \phi_0 | f(H_0 - E_0)f | \phi_0 \rangle = \langle \phi_0 | f [H_0, f] | \phi_0 \rangle. \tag{25}$$

if $H_0 \phi_0 = E_0 \phi_0$. If H_0 is of the form $-\alpha \nabla^2 + U$, with U any function of position, then it takes a bit of tedious algebra to work out the result

$$\langle \phi_0 | f [H_0, f] | \phi_0 \rangle = \alpha \langle \phi_0 | \mathrm{grad}\, f \cdot \mathrm{grad}\, f) | \phi_0 \rangle. \tag{26}$$

Since we obviously have $2\langle \phi_0 | fV_2 | \phi_0 \rangle$ as the leading term in (24), the whole functional now involves only expectation values over ϕ_0. Doing the angular integrals leaves us with radial expectation values which are obtainable from the power series method. One simple approximation is to set $f = kV_2$. For my $2p_0$ example this gives for the first term in the functional

$$2k \langle V_2^2 \rangle = 2k \langle r^4 \rangle \langle (3 - 5\mu^2)^2 \rangle \tag{27}$$

where I set $\lambda = \gamma^2/40$ and write $3x^2 + 3y^2 - 2z^2$ as

$$(3r^2 - 5z^2) = r^2(3 - 5\mu^2) \tag{28}$$

with $\mu = \cos\theta$ in polar coordinates. The expectation value is taken over a p_0 state, so the wavefunction has a factor μ. Using double brackets to denote averages over a sphere the angular expectation value becomes

$$\langle (3 - 5\mu^2)^2 \rangle \equiv \langle\!\langle \mu^2(3 - 5\mu^2)^2 \rangle\!\rangle / \langle\!\langle \mu^2 \rangle\!\rangle. \tag{29}$$

Using the standard result $\langle\!\langle \mu^n \rangle\!\rangle = (n + 1)^{-1}$ for even n leads to the result

$$2k \langle V_2^2 \rangle = 2k \langle r^4 \rangle \tfrac{12}{7}. \tag{30}$$

To work out the second term in the functional we need to work out $\mathrm{grad}\, V_2 \cdot \mathrm{grad}\, V_2$ which is just

$$\left(\frac{\partial V_2}{\partial x}\right)^2 + \left(\frac{\partial V_2}{\partial y}\right)^2 + \left(\frac{\partial V_2}{\partial z}\right)^2 = r^2(36 - 20\mu^2). \tag{31}$$

The result for the second term of the functional is

$$k^2 \langle V_2 [H_0, V_2] \rangle = \tfrac{1}{2} k^2 \langle r^2 \rangle 24. \tag{32}$$

The minimum of a function $2Ak - Bk^2$ is $A^2 B^{-1}$. Taking A and B from the results above gives an estimate for the second-order energy coefficient E_2. Since the perturbation coefficient, called λ in §9.2, is actually $\gamma^2/40$ here, the total second-order effect turns out to be

$$-0.2449 \left(\frac{\gamma^2}{40} \right)^2 \frac{\langle r^4 \rangle^2}{\langle r^2 \rangle}. \tag{33}$$

Doing the calculation at $\gamma = 0.1$ for the $2p_0$ state I obtained the energy $-0.111\,753$ using the series method program and also found the expectation values $\langle r^2 \rangle = 24.0078$, $\langle r^4 \rangle = 995.987$. The second-order shift due to V_2 is thus estimated to be $0.000\,63$, giving a corrected energy of $-0.112\,383$. This will still be slightly high, since the E_2 estimate could be slightly improved with a better trial function. Praddaude [9], using a large scale matrix calculation, obtained the result $-0.112\,410$ for this case, so I have done quite well using the power series method and a little perturbation theory!

The unperturbed $2p_0$ energy is -0.125, so the V_1 part of the perturbation has given an energy shift $+0.013\,247$, while V_2 has given a shift $-0.000\,63$. These two shifts are in the ratio $21:-1$. From a matrix-variational point of view the dropping of V_2 from the Hamiltonian is equivalent to setting up the matrix of the Hamiltonian in a basis of states which all have $l = 1$, $m = 0$. The resulting matrix energy value cannot be lower than my result $-0.111\,753$ (at $\gamma = 0.1$) because I have found the eigenvalue 'exactly', which corresponds to using a complete set of basis states of $l = 1$, $m = 0$ type. I have pointed out [8] that this provides a way of calibrating basis sets to judge their quality; for example, a basis of hydrogenic unperturbed bound state functions $2p_0$, $3p_0$, etc cannot work in principle and so must give an energy which is slightly high. The calculation described above can be carried out for various other states; for example the $3d_2$ state, although an excited state of the hydrogen atom, is actually the ground state for the family of states of even parity with $m = 2$. In my paper [8] I give numerical results for six states and also give a formula (from angular momentum theory) which gives the numerical factor A_1 to be used in forming V_1:

$$\tfrac{1}{8} \gamma^2 (x^2 + y^2) \equiv A_1 \gamma^2 r^2 + V_2 \tag{34}$$

$$A_1 = \tfrac{1}{12} \left[1 - \frac{3m^2 - l(l+1)}{(2l-1)(2l+3)} \right]. \tag{35}$$

12.6   Quasi-bound states

The simple finite-difference methods of §§10.2 and 10.6 use node counter and interpolation procedures to calculate the energies of bound states. Of course, not all states are bound states. It is well known that the hydrogen atom has a continuous spectrum of positive energy states, with wavefunctions which are not square integrable. (These functions plus the bound state ones are needed to give a complete set for the Zeeman effect problem §12.5.) In this section I shall study the Schrödinger equation

$$-\tfrac{1}{2}\nabla^2 \psi + \lambda r^2 \exp(-\beta r)\psi = E\psi \qquad (36)$$

which involves a spherically symmetric potential. For s states, with $\lambda > 0$ and $\beta = 0$ a harmonic oscillator problem results, giving regularly spaced bound state energies. With $\lambda > 0$ and $\beta < 0$ the potential rises even more rapidly with r, still giving only bound states. With $\lambda < 0$ and $\beta > 0$ the potential has a minimum at some r value and is zero at $r = 0$ and $r = \infty$. There will be some bound states (for which the particle is trapped in the well with negative total energy) and also some positive energy unbound states. The energy values can be found for any angular momentum l, but the finite-difference method (§10.6) is particularly interesting when we look at the case $l = 0$, $\lambda > 0$, $\beta > 0$. The potential looks like an oscillator one up to a distance of order β^{-1} but has a maximum and then falls to zero for $r \to \infty$. If the maximum potential is V_0 at $r = r_0$ then a classical particle with total energy less then V_0 would move around inside the sphere of radius r_0 without escaping. In quantum mechanics we are used to the idea of tunnelling; the particle will eventually get through the potential hill. In time-dependent quantum mechanics the use of an initial normalised wavefunction ψ_0 entirely within the central well will produce at later times a wavefunction which has a value less than 1 for P, the integral of ψ^2 over the inner region. Although P does not decay exactly according to an exponential law, a rough lifetime can be assigned for the probability decay process.

   In a time-independent approach any attempt to find a bound state energy should fail in principle, but many conventional calculations *do* appear to give some kind of result. For example, if the matrix of the energy operator is set up in a basis of oscillator states appropriate to $\beta = 0$, then for small β it will appear that the use of more basis states leads to a limiting energy, although this plateau value will eventually be lost as more and more basis states are used. This phenomenon of temporary stability is the basis of the so-called stabilisation method for dealing with these quasi-bound states [10, 11]. The point seems to be that as more basis states are added the region of r covered by the basis set is gradually expanded, and the stability indicates that over some range of r the imposition of the Dirichlet condition $\psi(r) = 0$ leads to an almost constant energy. For a true bound state, of course, an infinite range of r (right up to infinity) is involved. For several-particle problems the matrix form of the stabilisation method is useful, but for a one-dimensional problem it is much easier to get directly at the variation of E with r by using a finite-difference method or other technique which does not need the use of matrices or the intervention of a basis set. One widely used procedure is to set an r value, impose the condition $\psi(r) = 0$ and find the energy $E(r)$. As I pointed out when I looked at this problem [12] it is easier to study the inverse function $r(E)$, since a node counter instruction such as that in my program of §10.6 can easily be modified to print out the node positions for any assigned E. This is usually easier than doing a sequence of eigenvalue calculations as r is varied.

The finite-difference methods of §10.2 use some stripwidth h. If the ratio variable $R(r)$ is negative this means that $\phi(r)$ and $\phi(r + h)$ have opposite sign, and it might seem that the node position is only known to within a distance h. However, by using a little linear interpolation (or drawing some similar triangles) we can soon get a more accurate estimate of the node position:

$$X = r + h[1 - R(r)]^{-1} \tag{37}$$

(Set $R = -1, 0$, etc to see that it works). Even using this formula does not give us the exact node positions, since we need to have $h \to 0$ to get to the correct Schrödinger differential equation. As might be anticipated from all that has gone before, all that is needed is to use two small h values and a Richardson extrapolation to get very good node positions. Although I haven't used it here, it may be that a method such as Numerov's (§10.5) will give 'first time' node positions which are sufficiently accurate, when it is used in a form which permits the interpolation in equation (37) to be performed. For a potential such as that of equation (36) we have an essentially free particle at large r and the wavefunction will become oscillatory with a definite de Broglie wavelength. The distance between successive nodes will be constant, which provides a useful empirical way of spotting when the integration has proceeded to the tail of the potential. A function which varies as $\exp(ikr)$ has wavelength $2\pi k^{-1}$ and energy $k^2/2$. To convert a node position X into units of free particle wavelengths and then to multiply this by 2π to get an 'effective phase angle' η we can use the formula

$$\eta(E) = X\sqrt{2E}. \tag{38}$$

I give below a program which works out the node positions and η values for three energies (E and $E + DE$) at a time, thus allowing an estimate of the slope $(\partial \eta / \partial E)$ to be made when DE is small. The task of finding the E value at which the slope is a maximum is an interpolation problem of the kind discussed in §4.2. The result is almost identical if we use the criterion that $(\partial X / \partial E)$ shall be a maximum; this would be in keeping with the idea of the stabilisation method, since it would correspond to having E as stable as possible while X varies. The η criterion is more in the spirit of scattering theory, which studies phase shifts and describes a resonant state energy E_R as one which contributes a term of form $-\tan^{-1}(\frac{1}{2}\Gamma/(E-E_R))$ to $\eta(E)$. Here is the program.

```
10  DIM R(2), F(2), Q(2)
20  INPUT H, E, DE, L
30  N = L : H2 = H * H : L = L + 1
40  FOR M = 0 TO 2 : R(M) = 1: NEXT M
50  N = N + 1: X = H * N
60  V = 7.5 * X * X * EXP(−X)
70  FOR M = 0 TO 2 : F = E + (M − 1) * DE
80  F(M) = (2 * N * (V − F)
            + F(M) * (N − L)/R(M))/(N + L)
90  R(M) = 1 + H2 * F(M)
100 IF R(M) < 0 THEN 120
110 NEXT M
115 GOTO 50
120 Y = X + H/(1 − R(M)) : Y = Y * SQR(2 * F)
130 Q(M) = Q(M) + 1
140 PRINT M, Q(M), Y : GOTO 110
```

The reader may check (as I have done) that the quantities X, η, etc obtained using a stripwidth h obey the usual error laws which we have encountered before. Richardson extrapolation based on the results for $h = 0.01$ and $h = 0.02$ gives good results. The test equation (36) with $\beta = 1$, $\lambda = 7.5$, $l = 0$ has been treated by various authors (see [12]) so I quote some results which I obtained using the program given above. The results are for the ninth node, which is well out in the region where the wavefunction has settled down to a constant wavelength.

Energy	X	η	$\Delta\eta$
3.424	14.136 14	36.992 44	
			7672
3.425	14.104 77	36.915 72	
			7797
3.426	14.072 93	36.837 74	
			7835
3.427	14.040 95	36.759 40	
			7779
3.428	14.009 20	36.681 61	
			7628
3.429	13.978 02	36.605 33	

From the results for the differences $\Delta\eta$ it is clear that the maximum of $(d\eta/dE)$ appears at about 3.4265. Taking further differences of the last column we can form the Newton-Gregory interpolation series (§4.2).

$$\Delta\eta = 7797 + 38x − 47x(x − 1) \tag{39}$$

and from this we find that the value of $\Delta\eta$ has a maximum of 0.078 35 at $x = 0.90$, i.e. $E = 3.4264$. The $\Delta\eta$ value corresponds to a $(\mathrm{d}\eta/\mathrm{d}E)$ value of 7.835. If we follow the scattering theory idea that a resonant state should give a contribution to η described by the function $-\tan^{-1}(\frac{1}{2}\Gamma/(E-E_R))$, it follows after some algebra [10] that the value of the parameter Γ is given by twice the reciprocal of the maximum $(\mathrm{d}\eta/\mathrm{d}E)$ value. This yields $\Gamma = 0.0255$ from the results above. The potential has a maximum value $V_0 = 4.06$ at $r \sim 2$, so the energy 3.4264 is below this maximum. By changing the potential to $7.5r^2$ the reader may check that the lowest bound state is at $E \cong 5.81$. Using the potential with $\beta = 1$ but with the condition $\phi(2) = 0$ lowers the energy to 3.59. Keeping the potential constant beyond $r = 2$ yields a proper bound state with energy $E \cong 3.43$. The procedure for this last calculation is very simple and arises from the way in which the variables are treated in my radial equation program of §10.6. All that is required is to put after the statement which works out $X = N * H$ in that program the statement

IF X $>$ 2 THEN LET X $=$ 2

This holds the *potential* at that for $r = 2$ while letting all the other quantities advance properly along the axis. (Look at it and see; if you don't agree with me that it is beautiful, then how did you manage to get so far through this book?) Putting the proper tail on the potential instead of letting it continue at the peak value V_0 thus has hardly any effect on the E value but introduces the leakage effect characterised by a non-zero Γ parameter.

I have looked at one simple method for treating quasi-bound states which is closely related to the bound state calculational methods described earlier. There are other approaches. For example, the rotated coordinate approach uses a complex scaling parameter to convert the Schrödinger equation to another one which has complex energy eigenvalues, Γ being related to the imaginary part of the energy [13]. The least squares approach [14] varies a normalised trial function ψ so as to minimise the quantity $\langle\psi|H^2|\psi\rangle - \langle\psi|H|\psi\rangle^2$. For a true bound state this quantity would be zero, whereas the best that can be done is to give it a minimum value δ, say, for a quasi-bound state. The $\langle\psi|H|\psi\rangle$ value is taken to give E. My personal view is that Γ should be derivable from δ, but I haven't seen this done so far. In my paper [14] I relate the least squares approach via the iterative inverse calculation (§3.4) to the problem of looking for the E values at which the resolvent operator $(H-E)^{-1}$ has its poles. For a bound state the resolvent has a pole at the real E value, whereas for a quasi-bound state it has a pole at $E - \mathrm{i}(\Gamma/2)$ where Γ is the parameter which appeared in the η calculation above. The reason why I called this section 'quasi-bound states' instead of 'resonant states' is that for the example which I gave the various extra calculations which I carried out show that the state really *is* an almost-bound state. *My* interpretation of the η calculation is as follows. If η varies very rapidly with E it follows that by forming a wavepacket of functions $\psi(E)$ with E values in the region $E \pm \Gamma$ we can arrange to get strong destructive interference in the outer region (with a wide range of phases). In the inner region all the $\psi(E)$ are very similar, so we get strong constructive interference. The resulting wavepacket is strongly peaked inside the barrier and has an energy expectation value $\langle \psi | H | \psi \rangle$ close to E, with a value for $\langle \psi | H^2 | \psi \rangle - \langle \psi | H | \psi \rangle^2$ of order Γ^2. This is then a type of quasi-bound state, with a decay lifetime proportional to Γ^{-1}. Scattering theorists often describe a resonance energy as one for which it is possible to form such a wavepacket with a dominant inner portion [15]. The least squares and resolvent approaches are presumably also finding properties of some kind of optimum wavepacket quasi-bound state. It is interesting to see that even the simple methods of §10.6 can be modified to give results for quasi-bound states, leading to an estimate of Γ from a real variable calculation. One feature which the reader should be able to find out for himself by calculation is that the lifetime parameter Γ^{-1} depends very markedly on λ and β. What we have here is a simple 'radioactivity' problem of the type discussed qualitatively in textbooks of nuclear physics, except that the nuclear potential involved in radioactive decay is usually envisaged as a square well potential with a Coulomb type repulsive potential tail at large distances. The kind of η calculation done above could be done for such a potential, but it is clear that the estimated lifetime can be varied over many orders of magnitude by making a small change in the parameters which describe the potential. Thus, while the existence of radioactivity is qualitatively allowed by quantum mechanics, a detailed calculation of a radioactive lifetime would be very difficult. Quantum mechanics similarly allows qualitatively the existence of the chemical bond, but the calculation of the bond energies in a several-electron molecule is a difficult numerical task.

Notes

1. W Conley 1981 *Optimization: A Simplified Approach* (New York: Petrocelli, Conley)
2. G Relf March 1982 *Practical Computing* p 93
3. J Killingbeck 1972 *Mol. Phys.* **23** 913
4. R P Hurst, J D Gray, G H Brigman and F A Matsen 1958 *Mol. Phys.* **1** 189
5. J Killingbeck 1973 *J. Phys. B: At. Mol. Phys.* **6** 1376
6. J Killingbeck 1975 *J. Phys. B: At. Mol. Phys.* **8** 1585
7. J Killingbeck and S. Galicia 1980 *J. Phys. A: Math. Gen.* **13** 3419
8. J Killingbeck 1981 *J. Phys. B: At. Mol. Phys.* **14** L461
9. H C Praddaude 1972 *Phys. Rev.* **6** A 1321
10. A U Hazi and H S Taylor 1970 *Phys. Rev.* **1** A 1109
11. A Macias and A Riera 1980 *J. Phys. B: At. Mol. Phys.* **13** L449

12. J Killingbeck 1980 *Phys. Lett.* **77** A 230
13. W P Reinhardt 1976 *Int. J. Quant. Chem. Symp.* No 10 359
14. J Killingbeck 1978 *Phys. Lett.* **65**A 180
15. R D Levine 1969 *Quantum Mechanics of Molecular Rate Processes* (Oxford: Oxford Universities Press)

13 Some recent developments

13.1 Introduction

In this chapter I bring the material of the book up to date in three respects. First, I describe the ways in which I and other workers have improved and extended the methods described in the first edition of the book. Second, I describe the features of a few of the currently available microcomputers, with emphasis on the kind of facilities relevant for numerical work. Third, I give details of a new calculational method which seems to be of promise for both numerical and perturbation theoretic calculation of Schrödinger equation eigenvalues.

13.2 Recent mathematical work

The power series method (Chapter 7) and the finite difference method (Chapter 10) both lead to programs which give accurate eigenvalues fairly quickly, without keeping a record of the wavefunction. The eigenvalue differencing approach of §9.4 allows expectation values to be obtained, but this requires a sequence of energy calculations. I have recently discovered [1] how to combine the calculations in one simple program, so that an energy eigenvalue and some selected expectation value are produced together, both accurate to the same number of digits.

The hypervirial perturbation method (Chapter 9) has been applied to many different problems in the research journals [2, 3, 4]. I have noticed in my own work with the method that problems involving strong perturbations or excited states can cause computer overflow in a program such as that of §9.7. This overflow arises because the intermediate $B(N, M)$ coefficients can be much larger than the energy coefficient $E(M)$ which they combine to produce. The problem is much alleviated by introducing a scaling parameter F, typically 8 or 16 or some number which translates perfectly from decimal to binary.

The coefficients appearing in equations (36) and (37) of §9.7 are then modified
by the replacement

$$A(N, M) \rightarrow F^N A(N, M) = B(N + 2, M + 1)$$

and the resulting modified version of equation (36) is divided throughout by
F^N. The modified program involves a few extra coefficients involving low
powers of F, but is much less subject to overflow. Further, by adding a term
$\frac{1}{2}l(l + 1)r^{-2}$ into the potential it is possible to extend the perturbed Coulomb
potential program of Appendix 2 to apply to states of arbitrary angular
momentum. I have now changed Appendix 2 to show the program for arbitrary
l, as run on a Sinclair Spectrum microcomputer, and including the F factor to
cut down overflow problems.

The Hill determinant method (§8.3) has been the cause of some controversy
in the literature [5, 6]. I played some role in this dispute as an adjudicator, and
made a comment then which still stands: the most simple way to show that the
method can lead to spurious eigenvalues would be to exhibit one. So far I have
not seen one. Hautot and Nicolas [7] have produced a version of the Hill determinant method which incorporates convergence factors rather similar to those
used in the methods of Chapter 7 of this book. Bhattacharjee *et al* [8] have
described a Hill determinant method for calculating the bond energy and length
of the H_2^+ molecular ion.

There have been many recent papers and books on finite difference and
finite element methods, and so I can mention only a few. Guardiola and Ros [9]
gave a summary of various finite difference methods for the Schrödinger
equation. Grinstein *et al* [10] described a multi-grid approach for the two-
dimensional Schrödinger equation. Bender [11] gave simple examples of the
use of finite elements in quantum mechanics, with particular reference to the
solution of operator equations of motion. Moncrief [12] and Bender and Sharp
[13] also treated operator equations of motion, showing that some finite difference approaches rigorously preserve required properties of unitarity. For the
equations of motion of classical mechanics, Greenspan [14] and Kanatani [15]
have described finite difference methods which satisfy energy conservation
requirements. I found two of the recent books on finite elements to be particularly interesting. The books by Livesey [16] and by Silvester and Ferrari [17]
give clear accounts of the Rayleigh–Ritz theory in its finite element form.
Silvester and Ferrari describe finite element methods in electrical field theory,
which, since they involve the Laplacian operator, will also be of interest in
connection with the Schrödinger equation. Shampine [18] has described
methods for solving systems of ordinary differential equations which use several
step lengths and Richardson extrapolation ideas. Finite difference and finite
element methods for the study of solitons were surveyed by Michell and

Schoombie [19]. The use of splines in the solution of differential equations has been treated by several authors [20, 21]. Jamieson [22] discussed some difficulties in the finite difference approach to singular potential problems. I have recently devised an intriguing series method (involving divergent series) for one of his test problems [23].

The methods of this book in the main avoid the use of matrices for the one-dimensional Schrödinger equation, so I was interested to see some recent works which discussed the matrix variational approach and its limitations. Dyall *et al* [24] looked at the effects which arise when approximate finite matrix representations are used for quantum mechanical operators. Klanh and Morgan [25] showed that it is possible for some expectation values to diverge even while the energy converges as more basis functions are used in a matrix variational treatment of the Schrödinger equation.

13.3   Comments on selected microcomputers

My main interest in microcomputers is centred on their mathematical capabilities, rather than their graphics or networking facilities. I comment below on some machines which I have found to have useful features as far as the calculations of this book are concerned.

The BBC B computer

I have used the BBC computer from time to time in my visiting lectures at various universities and so have tried out several of my calculations on it. It uses the 6502 chip and is subject to the same kind of slight subtraction error which I describe for the Pet in §2.2. Array subscripts start at 0 in BBC BASIC and the word LET is not required in statements. This means that the Pet programs throughout this book are well suited for use on the BBC machine, which will actually run them more quickly. The BBC language has a few unexpected features, two of which I can illustrate as follows. If the variable Y has not appeared previously in a program the statement

$$Y = Y + (3 > 2)$$

will assign a value -1 to the new variable Y. The Y on the right is assigned a zero value and the expression $(3 > 2)$ is assigned the Boolean value for 'true', which is -1 on the BBC computer. A feature which I think has some potential for some calculations is that provided by the SPOOL and EXEC commands, which will allow two stored programs to be joined together, if care is taken to use different line numbers in the programs. This would permit the use of a standard matrix eigenvalue routine (with line numbers 200 onwards) with a variety of routines which calculate the matrix elements for particular problems (with line numbers less than 200). Other features of BBC BASIC are the REPEAT-UNTIL and IF-THEN-ELSE structures and the use of procedures. A procedure is similar in operation to a subroutine, but is called by its name rather than by a line number. The last three features are of the kind recommended by the proponents of structured programming; I have discussed this in one section of a recent review article [23]. BBC BASIC allows a variable to be given a local value within a procedure without disturbing its value as set in the main program.

The Tandy TRS 80

This older machine, with Tandy Level II BASIC, is still widely available on the second-hand market. I have found it useful for some of my calculations, since it has a DEFDBL instruction which permits the use of double precision variables and performs the four standard arithmetic operations with 16 decimal digits. The Level II BASIC also has several of the BBC BASIC features, such as the IF-THEN-ELSE structure and the AUTO and TRACE instructions which simplify program writing and debugging.

The Jupiter Ace

The high level language FORTH was originally used as a control language. Although it is speedy compared with BASIC, most of its implementations use integer arithmetic. However, the Jupiter Ace microcomputer [26] has FORTH as its built-in language and can do floating point arithmetic of 6 decimal digit accuracy. Currently the Ace (for which a 16K RAM pack is obtainable) is one of the lowest priced machines available, and makes it possible to learn the FORTH language while at the same time doing some real calculations. The arithmetic operations in FORTH use a stack, so the relevant language instructions are in a reverse Polish notation. For example the instructions 3.2 7.1 $F + F$. will add 3.2 to 7.1 and display the result 10.3, removing it from the top of the stack. The basic procedure in FORTH is to define new words (essentially operators) which will act on input numbers to produce any desired function of them. I have satisfactorily tried out some of the calculational methods of Chapter 10 on the Ace and in my microcomputer review article [23] I give a lengthy example of a calculation involving loops in Jupiter Ace FORTH.

The Sinclair Spectrum

The Spectrum Plus, with a new-style keyboard, is now available, although in my own work I have found it quite satisfactory to equip the original Spectrum with a DKtronics keyboard when it is to be given repeated use in teaching classes. Although the QL computer is currently the largest one in the Sinclair range, the Spectrum is still very useful for scientific work. Indeed, it is my experience tht many pupils and students have Spectrums at home which they could use (given the right books and software) to help their learning of mathematics and physics, particularly at A-level and at degree level. Although the more expensive BBC machine is ensconsed in many schools, it does not yet seem to have been properly integrated into the teaching of A-level mathematics and science (as opposed to computer studies).

Although the Spectrum apparently displays only 8 digits, it actually calculates to the same accuracy as many other machines. If at the end of a calculation the Spectrum shows E as 1.060362 1, for example, the manual instruction PRINT E-1.0 will reveal further digits. For calculations not involving too many arithmetic operations I find that the ninth digit is reliable. It is sometimes said that the Spectrum keyboard, with its unorthodox single key entry, it not suitable for serious computing. My view, on the contrary, is that it only yields its full benefits when it *is* used seriously. It just needs a little practice to reach the 'turn over point' at which the Spectrum keyboard can be used more quickly than a conventional one. In my work, as this book makes clear, I tend to get the mathematics right first, so that the programming structures only involve a few dozen standard BASIC instructions. The Spectrum keyboard is quite efficient for such work.

As far as numerical work is concerned the Spectrum BASIC includes several useful features. A numerical input can take the form of an expression e.g. SQR2 or 2/3. It is even possible to input a symbol as a numeric input, if the symbol is the name of a variable which already has a value. (Most other computers will not allow this, since it looks as though a string is being used instead of a number.) For example, suppose that four variables A, B, C and D have been given decimal values during a program run and that, either automatically or manually, we GO TO the input line for a new set of four values. If the operator wants to change D only he can input the first three variables as 'A', 'B' and 'C' and only put in D explicitly as a full decimal number. That saves a lot of effort. As far as user defined functions are concerned the DEF FN operation can be used to define functions of one or two variables at a time. Alternatively, the operator can input a string variable such as $X * X + 3$. A request at any later stage for the value of that string variable (using VAL) will yield the number $X^2 + 3$ if X has a numerical value. This facility makes it possible to change functions in a program easily without writing in program lines using DEF FN. Boolean functions can also be used in numerical calculations; they yield 1 for true and 0 for false. I have used them to count nodes in a wavefunction and also to simulate a delta function in a finite difference calculation [1].

Although the Spectrum normally operates in BASIC, it can also be used to explore the value of other high level languages for scientific work. A PASCAL compiler [27] is available which gives 7 digit floating point arithmetic; there is also a FORTH compiler [28] which gives the same accuracy as the usual Spectrum BASIC.

There are many popular magazines dealing with the Spectrum and other microcomputers, but I find that the weekly series *Input*, published by Marshall Cavendish, stands out as particularly informative. It contains sections on BASIC and on machine code and it is carefully and clearly written. Each program is described in detail, with versions for the Spectrum, BBC, Tandy and CBM microcomputers. I have found the series particularly useful in my recent attempts to mix in graphics and machine code with my BASIC programs. Two recent books which are useful for microcomputer mathematical work are the one by Mason [29], which looks at a range of numerical methods (with BASIC programs), and the one by Kantaris and Howden [30], which studies several mathematical problems which involve finding the roots of equations.

13.4   The inner product method

The diagonal hypervirial recurrence relations involve expectation values such as $\langle \psi | x^N | \psi \rangle$, where ψ is some unknown eigenfunction of the Schrödinger equation. It is also possible to derive recurrence relations between the inner product quantities

$$S(N) = \langle \phi | x^N | \psi \rangle \tag{1}$$

where ϕ is an appropriately chosen reference function. As an illustrative example I will use the perturbed oscillator Schrödinger equation

$$-D^2 \psi + x^2 \psi + \lambda x^4 \psi = E \tag{2}$$

with the reference function

$$\phi = x^P \exp(-\beta x^2 / 2). \tag{3}$$

In (3) the parity index P is 0 for even parity states and 1 for odd parity states. The eigenfunctions ψ of (2) are real and the Hamiltonian H is Hermitian, so that the identity

$$\langle \phi | x^N H | \psi \rangle = \langle \psi | H x^N | \phi \rangle \tag{4}$$

will hold. Both the left-hand and right-hand sides of (4) can be worked out because of the simple choice of ϕ. The result is a recurrence relation involving the $S(N)$:

$$[E - (2N + 2P + 1)\beta] S(N) + N(N + 2P - 1) S(N - 2)$$
$$= (\mu - \beta^2) S(N + 2) + \lambda S(N + 4). \tag{5}$$

Introducing the variables $R(N)$ through the defining equation

$$S(N + 2) = R(N)S(N) \tag{6}$$

gives the result

$$R(N - 2) = N(N + 2P - 1)/T(N) \tag{7}$$

where

$$T(N) = (2N + P - 1)\beta - E + (\mu - \beta^2)R(N) + \lambda R(N)R(N + 2). \tag{8}$$

For the special case $N = 0$ equation (5) can be written as

$$E = (2P + 1)\beta + (\mu - \beta^2)R(0) + \lambda R(0)R(2). \tag{9}$$

The computational use of (8) and (9) is as follows. For some large N value, $N = Q$, the quantities $R(Q)$ and $R(Q + 2)$ are set equal to zero. (This is trivial, since the Spectrum and many other microcomputers set array elements at zero once the array has been declared.) Then a trial E is used and the $R(N)$ down to $R(0)$ are evaluated using (8). E is then computed using (9). If the input and output E values agree then E is an eigenvalue. In principle the eigenvalues depend on β and Q, but as Q increases they settle down to limiting values which are eigenvalues of the Schrödinger equation (2) [31].

The program given on p 171 computes the quantity $E(\text{output}) - E(\text{input})$ as a function of E, in steps of 0.2, and then homes in on an eigenvalue by using a modified Newton's method similar to that given in Chapter 3. The reader may check that it gives the following test results with $\beta = 5$, $\mu = 1$, $\lambda = 1$, and $Q = 100$.

Even parity $E = 1.3923516,$ 8.6550500,   18.057557

Odd parity $E = 4.6488127,$ 13.156804,   23.297441.

One interesting feature of this inner product method is that it can also take a perturbation theoretic form. If the quantities E and $S(N)$ in equations (5) and (6) are postulated to be power series in λ, then recurrence relations are obtained which can be solved to give the energy perturbation series at least for the ground state of each parity. I regard this form of the theory as of great promise. Austin [32] has used it to get some renormalised energy perturbation series for the hydrogen atom Zeeman problem, outlined in §12.5, although that problem still seems to defy the hypervirial approach. M N Jones and I have recently applied the inner product approach in a renormalised perturbation form to get reasonable eigenvalues for the non-separable Schrödinger equation

$$-\nabla^2 \psi + (x^2 + y^2)\psi + \lambda x^2 y^2 \psi = E\psi. \tag{10}$$

The program for the Schrödinger equation (2) is as follows for the Spectrum and readers will be able to modify it with little difficulty for other machines. (Note that array indices start from 1 on the Spectrum.)

The inner product program (Spectrum version)

```
 5 INPUT "dim, mu, la"; q, mu, la
10 PRINT "p, 0 or 1": INPUT p:
LET p = 2 * p
15 DIM r(q + 12)
18 LET h = .001: LET k = 1: LET ep = 0:
LET lim = .2
20 PRINT "beta": INPUT b: LET
par = mu − b * b
21 INPUT "e"; e
22 LET e = e − lim: LET y = ep: GO SUB 35:
PRINT e, ep: IF y/ep > 0 THEN
GO TO 22
26 GO SUB 35: LET y = ep
27 LET e = e + h: GO SUB 35: LET yy = ep
28 LET sh = y * h/(y − yy): IF ABS sh > ABS
lim THEN LET sh = lim * SGN sh
29 PRINT e − h, y: LET k = 1: LET e
= e − h + sh: GO TO 26
35 FOR n = q TO 2 STEP −2
40 LET t = (n + n + 1 + p) * b − e + r(n + 1) *
(par + la * r(n + 3))
45 LET r(n − 1) = n * (n + p − 1)/t: NEXT n
50 LET s = r(1) * (par + la * r(3)): LET
ep = s + b * (1 + p) − e
55 RETURN
```

Notes

1. J Killingbeck 1985 *J. Phys. A: Math. Gen.* **18** 245
2. E J Austin 1984 *J. Phys. A: Math. Gen.* **17** 367
3. F M Fernandez and E A Castro 1982 *J. Mol. Spectrosc.* **94** 28
4. A Maluendes, F M Fernandez, A M Meson and E A Castro 1984 *Phys. Rev.* A **30** 2227
5. G P Flessas and G S Anagnostatos 1982 *J. Phys. A: Math. Gen.* **15** L537
6. A Hautot and M Nicolas 1983 *J. Phys. A: Math. Gen.* **16** 2953
7. A Hautot and A Magnus 1979 *J. Comput. Appl. Math.* **5** 3
8. R S Bhattacharjee, R P Saxena, P K Srivastava and K V Sane 1983 *Phys. Rev.* A **28** 2042
9. R Guardiola and J Ros 1982 *J. Comput. Phys.* **45** 374
10. F F Grinstein, H Rabitz and A Askar 1983 *J. Comput. Phys.* **51** 423
11. C M Bender 1984 *Physica* A **124** 91
12. V Moncrief 1983 *Phys. Rev.* D **28** 2485
13. C M Bender and D H Sharp 1983 *Phys. Rev. Lett.* **50** 1535
14. D Greenspan 1984 *J. Comput. Phys.* **56** 28
15. K I Kanatani 1984 *J. Comput. Phys.* **53** 181
16. R K Livesey 1983 *Finite Elements: an Introduction for Engineers* (Cambridge: Cambridge University Press)
17. R P Silvester and R L Ferrari 1983 *Finite Elements for Electrical Engineers* (Cambridge: Cambridge University Press)
18. L F Shampine 1983 *I.M.A. J. Num. Anal.* **3** 383
19. A R Mitchell and S W Schoombie 1984 *Numerical Methods in Coupled Systems* ed R W Lewis, P Bettess and E Hinton (New York: Wiley) ch 16
20. A K A Khalifa and J C Eilbeck 1982 *I.M.A. J. Num. Anal.* **2** 111
21. P Wang and R Kahawita 1983 *Int. J. Comput. Math.* **13** 271
22. M J Jamieson 1983 *J. Phys. B: At. Mol. Phys.* **16** L391
23. J P Killingbeck 1985 *Rep. Prog. Phys.* **48** 53
24. K G Dyall, I P Grant and S Wilson 1984 *J. Phys. B: At. Mol. Phys.* **17** 493
25. B Klanh and J D Morgan 1984 *J. Chem. Phys.* **81** 410
26. Jupiter Ace, supplier Boldfield Limiting Computing, Sussex House, Hobson Street, Cambridge CB1 1NJ
27. Spectrum PASCAL, supplier HiSoft, 180 High Street, North Dunstable, Beds LU6 1AT
28. Spectrum FORTH, supplier C P Software, 17 Orchard Lane, Prestwood, Great Missenden, Bucks HP16 0NN
29. J C Mason 1983 *BASIC Numerical Mathematics* (London: Butterworth)
30. N Kantaris and P F Howden 1983 *The Universal Equation Solver* (Wilmslow: Sigma Technical Press)
31. J Killingbeck, M N Jones and M J Thompson 1985 *J. Phys. A: Math. Gen.* **18** 793
32. E J Austin 1985 *Int. J. Quantum Chem.* to be published

Appendix 1

Useful mathematical identities

We take the angular momentum operators in the form

$$l_z = i\left(x\,\frac{d}{dy} - y\,\frac{d}{dx}\right) \tag{1}$$

and denote the total squared angular momentum by l^2. The Laplacian operator is denoted by the usual symbol ∇^2 and $\phi(r)$ and $f(r)$ are spherically symmetric functions. A solid harmonic Y_L is a function of (x, y, z) or (r, θ, ϕ) which obeys the equations

$$\mathbf{r} \cdot \operatorname{grad} Y_L = L Y_L : \nabla^2 Y_L = 0 \tag{2}$$

L is called the degree of the harmonic. Our useful identities can now be listed as follows.

1. $$l\left[F\phi(r)\right] = \phi(r)\left[lF\right] \tag{3}$$

   i.e. spherically symmetric factors act like constants for angular momentum operators.

2. $$l^2 F_L = \left[L(L+1) - r^2 \nabla^2\right]F_L \tag{4}$$

   if F_L is homogeneous of degree L in (x, y, z) i.e. if it obeys the *first* defining equation for a solid harmonic. If it obeys the second equation also, then the identity shows that Y_L is an angular momentum eigenfunction.

3. If the function ψ is a bound state wavefunction for the Schrödinger equation

   $$H\psi = -\alpha\nabla^2\psi + U(r)\psi = E\psi \tag{5}$$

   then the following expectation values with respect to ψ are equal to

   $$\langle f(\mathbf{r})\,[H, g(\mathbf{r})]\rangle = \alpha\langle \operatorname{grad} f \cdot \operatorname{grad} g\rangle. \tag{6}$$

Here [ ] denotes an operator commutator, and U, g and f are functions (not necessarily spherically symmetric) of (x, y, z). Also

$$\langle -\nabla^2 \rangle = \int \text{grad } \psi^* \cdot \text{grad } \psi \, dV. \tag{7}$$

These two results are useful in many calculations. The second one allows kinetic energy calculations to be performed for functions with discontinuous slope. My own reading suggests that the early workers in quantum mechanics defined the right-hand side to be the kinetic energy expectation value since it visibly has the semi-classical property of being always positive. In modern times the quantities $|\text{grad } \psi|^2$ and $U|\psi|^2$ are sometimes used as local kinetic and potential energy densities in discussions of the energy changes arising when atoms form a chemical bond.

4. $$\nabla^2 [Y_L e^{-f(r)}] = T Y_L e^{-f(r)} \tag{8}$$

where

$$T = f'^2 - f'' - 2(L + 1)r^{-1}f'. \tag{9}$$

This result is useful for providing test examples of Schrödinger equations with exactly known solutions. To get the one-dimensional case we formally set $L = -1$ in the above expression for T; a similar trick works with most formulae derived for three dimensions.

5. Consider two points with vector positions $\mathbf{r}$ and $\mathbf{R}(r < R)$ relative to the origin. Denote the angle between $\mathbf{r}$ and $\mathbf{R}$ by θ. Then we have the standard expansion (with $\mu = \cos \theta$ and $x = rR^{-1}$):

$$\frac{1}{r_{12}} = R^{-1} \sum x^k P_k(\mu) \tag{10}$$

where P_k is a Legendre polynomial and $r_{12} = |\mathbf{r} - \mathbf{R}|$. From the cosine rule we obviously have

$$r_{12}^2 = R^2(1 + x^2 - 2x\mu). \tag{11}$$

μ is just $P_1(\mu)$. Multiplying the above two formulae together and using the property $(1 + 2n)P_1 P_n = (n + 1)P_{n+1} + nP_{n-1}$ we find

$$r_{12} = R \sum_0^\infty P_k(\mu) \left[\frac{x^{k+2}}{2k + 3} - \frac{x^k}{2k - 1} \right]. \tag{12}$$

6. To work out the expectation value of the Coulomb repulsion operator r_{12}^{-1} for a product wavefunction $\psi(\mathbf{r}_1)\phi(\mathbf{r}_2)$ we have to integrate over *all* space and so must add *two* integrals. One involves the region of space with $r_1 > r_2$, the other refers to the region with $r_2 > r_1$. (The angular integrals

factor out as specific numbers.) The integral for $r_1 > r_2$ can be written as follows

$$I = \int_{r_1 > r_2} F(r_1)G(r_2)r_1^M r_2^N \, dr_1 \, dr_2. \tag{13}$$

(Here we may incorporate a factor $4\pi r^2$ in F and G to give a physical volume element if appropriate.) I can be worked out as a repeated integral,

$$I = \int_0^\infty y^M F(y) \left[\int_0^y x^N G(x) \, dx \right] dy \tag{14}$$

and can be handled analytically or numerically, depending on the form of F and G.

7. In atomic theory one-electron radial functions of type $r^n e^{-\alpha r}$ are often used, singly or in linear combinations, to represent the radial functions for each orbital. This means that repeated integrals of the type mentioned in 6 above have to be evaluated with the choice $F = e^{-\alpha r}$, $G = e^{-\beta r}$. Denoting the integral by $I(M, N; \alpha, \beta)$, it is possible to derive a very simple formula for the cases in which M and N are integers with $N \geqslant 0$ and $M + 1 \geqslant 0$.

$$I(M, N; \alpha, \beta) = \frac{M! \, N!}{\alpha^{K+1} r^{N+1} (1+r)^K} S(M+1, K) \tag{15}$$

where $r = \beta \alpha^{-1}$, $K = M + N + 1$, and $S(M+1, K)$ is the sum of the first $M + 1$ terms of the binomial expansion of $(1 + r)^K$, starting at the r^K term. The mathematical derivation of this simple computational prescription is given by Killingbeck [chapter 3, note 7]. For non-integer M and N a more complicated result involving sums of factorial function terms is obtained.

8. In one dimension the kinetic energy operator is a multiple of $-D^2$. In finite-difference methods it is approximated by some form of difference operator, which uses only discrete values of the wavefunction, e.g. $\psi(0)$, $\psi(h)$, $\psi(2h)$, etc for some step length h. It is sometimes useful to write the formulae of the continuous wavefunction theory to involve a step length h. For example, with ϵ a multiple of h,

$$\int \psi(x + \epsilon)\psi(x - \epsilon) \, dx = \int \psi^2(x) \, dx + \epsilon^2 T + 0(\epsilon^4) \tag{16}$$

where

$$T = \int [\psi\psi'' - \psi'^2] \, dx. \tag{17}$$

For a normalised ψ we see that T is just twice the integral of $\psi\psi''$ which means that we can compute the kinetic energy expectation value from integrals using ψ alone if we take ϵ sufficiently small. In practice we would play safe by using two ϵ values and a Richardson extrapolation (§4.3) to find T.

Appendix 2

The hypervirial program (Spectrum version)

In line 10 el is the angular momentum, q the principal quantum number, k the renormalising parameter and la the λ parameter.

```
   5  DIM b(30, 30) : DIM e(30)
  10  LET el = 0: LET q = 1: INPUT "k
, la"; k, la
  15  LET mu = 1 − k * la: LET e(1) = −mu *
mu/(2 * q * q)
  20  LET e = e(1): LET b(2, 2) = 1: L
ET m = 0: LET mf = 1
  22  PRINT e(1), m
  25  LET f = 32
  30  FOR n = 0 TO 25
  35  LET s = 0: LET nn = n + n
  40  FOR p = 0 TO m
  45  LET s = s + e(p + 1) * b(n + 2, m + 2 − p)
: NEXT p
  56  LET par = (n + 1) * (n − 1)/4 − el * (e
l + 1)
  58  LET s = (nn + 2) * s
  59  IF n > 0 THEN LET s = s + n * par * b
(n, m + 2)/(f * f)
  60  LET t = mu * b(n + 1, m + 2) + k * b(n + 1
, m + 1)
  65  LET s = s + (nn + 1) * t/f − (nn + 3) * b
```

```
   (n + 3, m + 1) * f
68 IF n = 0 THEN GO TO 72
70 LET b(n + 2, m + 2) = -(s/e(1))/(n
n + 2)
71 GO TO 75
72 LET b(1, m + 2) = -s * f/mu
75 NEXT n
80 LET z = b(3, m + 2) * f - k * b(1, m + 2)
/f
85 LET e(m + 2) = z/(m + 1)
87 LET mf = mf * la: LET ei = e(m + 2) *
mf: LET e = e + ei
88 PRINT ei, m + 1
89 IF m = 23 THEN STOP
90 LET m = m + 1: GO TO 30
```

Appendix 3

Two more Monte-Carlo calculations

1. An integration

One simple way to estimate a multiple integral, the integral of a function $f(x_1, x_2, \ldots x_N)$, is to choose the coordinates x_1 to x_N as a set of random numbers, work out f, and then repeat the process, computing an average f value from many runs. For example, the double integral of §5.8 has $f = \exp(-x - y)$ and the integration region $0 \leqslant x \leqslant 1$, $0 \leqslant y \leqslant 2$. Choosing x and y at random in these ranges (as explained in §12.3) gives the following results, each for 100 runs and each multiplied by 2, the physical volume of the region considered (to see why, think of the case $f = 1$). I quote only three figures.

$$0.535, \quad 0.491, \quad 0.548, \quad 0.568, \quad 0.544$$
$$0.566, \quad 0.555, \quad 0.542, \quad 0.517, \quad 0.508$$

Viewed as a statistical ensemble the results of these ten runs give a value of 0.537 ± 0.024, whereas the correct value of the integral is 0.5466 to four figures. For integrals in very many dimensions it can become quicker to use the Monte-Carlo approach, but it is clearly less effective for our simple two-dimensional example.

2. A Padé approximant

The $[1/2]$ approximant for the Euler series is

$$1 - z + 2z^2 - 6z^3 \ldots \quad = \quad (1 + 3z)(1 + 4z + 2z^2)^{-1} \tag{1}$$

as found by the result of §6.4. Using the variational principle, however, we use the trial function $(A + Bzt)\psi$ in the expression

$$2\langle\psi|\phi\rangle - \langle\psi|(1 + zt)|\psi\rangle \tag{2}$$

(see §6.3). Working out all the terms gives the result

$$\mu_0[2A - A^2] + \mu_1 z[2B - 2AB - A^2]$$
$$+ \mu_2 z^2[-B^2 - 2AB] + \mu_3 z^3[-B^2]. \tag{3}$$

For the Euler series we have $\mu_0 = \mu_1 = 1$, $\mu_2 = 2$, $\mu_3 = 6$. Varying A and B in a Monte-Carlo optimisation, as in §12.3, with $z = 0.1$, I obtained a maximum of $0.915\,492\,957$ at $A = 0.985\,88$, $B = -0.704\,11$. This agrees with the $[1/2]$ approximant at $z = 0.1$, and this agreement is also found at other z values. Now, the interesting point is that the identity holds for *any* series, whether or not a series of Stieltjes: it is the upper and lower bound properties (§6.3) which are the special property of Stieltjes series (although not exclusively). As an example we can just invent a function by expanding a rational fraction,

$$(1 + z)/(1 - z - 2z^2) = 1 + 2z + 4z^2 + 8z^3 + \dots . \tag{4}$$

This series is not a Stieltjes series, but has the *formal* coefficients $\mu_0 = 1$, $\mu_1 = -2$, $\mu_2 = 4$, $\mu_3 = -8$, when we compare it with the standard form of the series. Putting these values into the Monte-Carlo calculation at $z = 0.1$ gives a maximum of 1.25. This is the correct $[1/2]$ value, since the fraction given above is obviously the $[1/2]$ approximant to the series expansion on the right.

Appendix 4

Recurrence relations for special functions

Although I have not explicitly used them in my treatment of the Schrödinger equation, the special functions of Hermite, Laguerre and Legendre often appear as factors in either the wavefunction or the potential function when a traditional textbook approach is used. Other special functions appear in computing work or in connection with differential equations of theoretical physics. To calculate the numerical value of such a special function it is often preferable to use a recurrence relation rather than to use an explicit lengthy polynomial for the required function. Fox and Mayers [reference 1.7] devote a whole chapter to the use of recurrence relations and the problems to look out for when using them on a computer. Below I list some useful properties of several of the well known special functions of theoretical physics.

Legendre polynomials

$$P_n(x) = [2^n n!]^{-1} D^n [(x^2 - 1)^n] \qquad (|x| \leqslant 1)$$

$$(x^2 - 1)y'' + 2xy' - n(n + 1)y = 0$$

$$(n + 1)P_{n+1}(x) = (2n + 1)xP_n(x) - nP_{n-1}(x)$$

Hermite polynomials

$$H_n(x) = (-1)^n \exp(x^2) D^n [\exp(-x^2)]$$

$$y'' - 2xy' + 2ny = 0$$

$$H_{n+1}(x) = 2xH_n(x) - 2nH_{n-1}(x)$$

Laguerre polynomials

$$L_n(x) = \exp(x)D^n[x^n \exp(-x)]$$

$$xy'' + (1-x)y' + ny = 0$$

$$L_{n+1}(x) = (2n+1-x)L_n(x) - n^2 L_{n-1}(x)$$

Bessel functions

$$x^2 y'' + xy' + (x^2 - n^2)y = 0$$

$$xJ_{n+1}(x) = 2nJ_n(x) - xJ_{n-1}(x)$$

Chebyshev polynomials

$$T_n(x) = \cos[n \cos^{-1}x] \qquad (|x| \leqslant 1)$$

$$(1-x^2)y'' - xy' + n^2 y = 0$$

$$T_{n+1}(x) = 2xT_n(x) - T_{n-1}(x)$$

$$T_{nm}(x) = T_n(T_m(x))$$

In numerical analysis a function $f(x)$ is often represented as a linear combination $L(x)$ of Chebyshev polynomials when a representation is sought which gives a *minimum* value to the *maximum* value of $|f(x) - L(x)|$ in the range of x values considered. A book which discusses Chebyshev polynomials, Padé approximants and continued fractions is *Chebyshev Methods in Numerical Approximation* by M A Snyder 1966 (New York: Prentice-Hall).

Postscript:
some remaining problems

While writing this book I have noticed that there are several interesting problems which arise quite naturally out of the material which I have treated. The problems are essentially research projects: a satisfactory resolution of any one of them would probably lead to a publishable piece of work. I hand them over to any interested reader who wishes to develop further the methods which I describe in this book.

1. The off-diagonal problem

Using the ideas of §§9.4, 9.5 and 9.6 it is possible to calculate various expectation values for the eigenfunctions of the one-dimensional Schrödinger equation, using only energy calculations. Is it possible to reduce the calculation of at least some off-diagonal matrix elements (e.g. $\langle \psi_1 | r | \psi_2 \rangle$) between different eigenfunctions to a form which needs only energy computations? I suspect that off-diagonal hypervirial relations might be useful here: see reference 9.5.

2. Resonant state computations

In the calculation of the energies of resonant state energies (§12.6) some authors use the rotated Hamiltonian method (reference 12.13). This involves calculating complex eigenvalues with a complex potential, with the wavefunction decaying at infinity exactly as for a traditional real energy bound state. Can the methods of §§7.4 and 10.6 be modified to handle such cases?

3.   Monte-Carlo Padé approximants

The variational approach to Padé approximant calculation (§6.3, Appendix 3) does not appear to involve the division by zero problems which might arise in Wynn's algorithm. Can a Monte-Carlo approach to Padé approximant calculation be devised which is quick enough and accurate enough to compete with other procedures?

4.   Energy computations

Can any of the possible methods of eigenvalue estimation which I propose in §8.4 be developed for use on a microcomputer?

5.   Zeeman effect

In principle it is possible to compute the energy and wavefunction perturbation series for the hydrogen atom Zeeman effect (§12.5), but I give only the start of the 1s series. It would be interesting to see the series worked out for several states and then treated by a Padé approximant method. Might it be possible to get the hypervirial-perturbation approach to work for the problem? (That would be a tour-de-force!)

6.   Calculating the molecular energy

In calculating the molecular energy in §11.2 I found that the eigenvalue error is of order h^2 even when the choice $K = \frac{1}{12}h^2$ is made in the method of §10.3. This is (I think) because the potential is not smooth at $x = d$. It would be interesting to see whether the perturbation approach of §10.3 can be extended to apply to such cases.

Index